Digital Fashion Innovations

Digitalisation is becoming a standard practice in the fashion industry. Innovation in digital fashion is not just limited to computer-aided design (CAD) and manufacturing (CAM), rather it runs throughout the fashion supply chain, from product life cycle management and developing new business models that promote sustainability to connecting virtual and augmenting reality (VR/AR) with fashion for enhanced consumers experience through smart solutions. *Digital Fashion Innovations: Advances in Design, Simulation, and Industry* captures the state-of-the-art developments taking place in this multi-disciplinary field:

- Discusses digital fashion design and e-prototyping, including 2D/3D CAD, digital pattern cutting, virtual drape simulation, and fit analysis.
- Covers digital human modelling and VR/AR technology.
- Details digital fashion business and promotion, including application of e-tools for supply chain, e-commerce, block chain technologies, big data, and artificial intelligence (AI).

This interdisciplinary book will appeal to professionals working in textile and fashion technology, those developing AR and AI for clothing end uses, and anyone interested in the business of digital fashion and textile design. It will also be of interest to scientists and engineers working in anthropometry for a variety of disciplines, such as medical devices and ergonomics.

Textile Institute Professional Publications

Series Editor: Helen D. Rowe, The Textile Institute

Fibres to Smart Textiles: Advances in Manufacturing, Technologies, and Applications
Asis Patnaik and Sweta Patnaik

Flame Retardants for Textile Materials
Asim Kumar Roy Choudhury

Textile Design: Products and Processes
Michael Hann

Science in Design: Solidifying Design with Science and Technology
Tarun Grover and Mugdha Thareja

Textiles and Their Use in Microbial Protection: Focus on COVID-19 and Other Viruses
Jiri Militky, Aravin Prince Periyasamy, and Mohanapriya Venkataraman

Dressings for Advanced Wound Care
Sharon Lam Po Tang

Medical Textiles
Holly Morris and Richard Murray

Odour in Textiles: Generation and Control
G. Thilagavathi and R. Rathinamoorthy

Principles of Textile Printing
Asim Kumar Roy Choudhury

Solar Textiles: The Flexible Solution for Solar Power
John Wilson and Robert Mather

Digital Fashion Innovations: Advances in Design, Simulation, and Industry
Abu Sadat Muhammad Sayem

For more information about this series, please visit: www.routledge.com/Textile-Institute-Professional-Publications/book-series/TIPP

Digital Fashion Innovations

Advances in Design, Simulation, and Industry

Edited by Abu Sadat Muhammad Sayem

CRC Press is an imprint of the
Taylor & Francis Group, an **informa** business

First edition published 2023
by CRC Press
6000 Broken Sound Parkway NW, Suite 300, Boca Raton, FL 33487–2742

and by CRC Press
4 Park Square, Milton Park, Abingdon, Oxon, OX14 4RN

CRC Press is an imprint of Taylor & Francis Group, LLC

ISBN: 978-1-032-20729-2 (hbk)
ISBN: 978-1-032-20727-8 (pbk)
ISBN: 978-1-003-26495-8 (ebk)

DOI: 10.1201/9781003264958

Typeset in Times
by Apex CoVantage, LLC

Contents

PART C Digital Human and Metaverse

PART D Digital Business and Promotion

Series Preface

TEXTILE INSTITUTE PROFESSIONAL PUBLICATIONS

The aim of the ***Textile Institute Professional Publications*** is to provide support to textile professionals in their work and to help emerging professionals, such as final year or master's students, by providing the information needed to gain a sound understanding of key and emerging topics relating to textile, clothing and footwear technology, textile chemistry, materials science, and engineering. The books are written by experienced authors with expertise in the topic, and all texts are independently reviewed by textile professionals or textile academics.

The textile industry has a history of being both an innovator and an early adopter of a wide variety of technologies. There are textile businesses of some kind operating in all counties across the world. At any one time, there is an enormous breadth of sophistication in how such companies might function. In some places where the industry serves only its own local market, design, development, and production may continue to be based on traditional techniques, but companies that aspire to operate globally find themselves in an intensely competitive environment, some driven by the need to appeal to followers of fast-moving fashion, others by demands for high performance and unprecedented levels of reliability. Textile professionals working within such organisations are subjected to a continued pressing need to introduce new materials and technologies, not only to improve production efficiency and reduce costs, but also to enhance the attractiveness and performance of their existing products and to bring new products into being. As a consequence, textile academics and professionals find themselves having to continuously improve their understanding of a wide range of new materials and emerging technologies to keep pace with competitors.

The Textile Institute was formed in 1910 to provide professional support to textile practitioners and academics undertaking research and teaching in the field of textiles. The Institute quickly established itself as the professional body for textiles worldwide and now has individual and corporate members in over 80 countries. The Institute works to provide sources of reliable and up-to-date information to support textile professionals through its research journals, the ***Journal of the Textile Institute***[1] and ***Textile Progress***[2], definitive descriptions of textiles and their components through its online publication ***Textile Terms and Definitions***[3], and contextual treatments of important topics within the field of textiles in the form of self-contained books such as the ***Textile Institute Professional Publications***.

REFERENCES

www.tandfonline.com/action/journalInformation?show=aimsScope&journalCode=tjti20
www.tandfonline.com/action/journalInformation?show=aimsScope&journalCode=ttpr20
www.ttandd.org

Editor Biography

Dr. Abu Sadat Muhammad Sayem is the principal investigator of the UKRI project "Digital Fashion Network (2023-24)". He has been working in the field of digital fashion since 2004 and has been chairing the annual event "Digital Fashion Innovation Symposium" since 2020. His ongoing research at the Manchester Metropolitan University covers the areas of digital prototyping, sustainable materials and process innovation, and smart garments. He obtained a PhD from the University of Manchester, an MSc. from the Technische Universität Dresden and a BSc. from the University of Dhaka. In the past, Dr. Sayem worked as an Associate Professor in the Textile Department and as the Head of the Centre of Scientific Research and Innovation of the Southeast University, Bangladesh. He is a member of the UKRI Talent Peer Review College, a fellow and trustee of the Textile Institute, a fellow of the Higher Education Academy (Advance HE, UK) as well as an alumnus of the Commonwealth Scholarship Commission, UK and the Deutscher Akademischer Austauschdienst (DAAD), Germany.

Contributors

Mominul Ahsan
Department of Computer Science
University of York
York, UK

Fatma Baytar
Department of Human Centered Design
Cornell University
Ithaca, New York

Samit Chakraborty
Wilson College of Textiles
North Carolina State University
Raleigh, North Carolina

Vien Cheung
3D Weaving Innovation Centre, School of Design
University of Leeds
Leeds, UK

Hilde Heim
Manchester Fashion Institute
Manchester Metropolitan University
Manchester, UK

S M Azizul Hoque
Wilson College of Textiles
North Carolina State University
Raleigh, North Carolina

Md. Mazharul Islam
Department of Textile Engineering
Northern University Bangladesh
Dhaka, Bangladesh

Anke Klepser
Hohenstein Institut für Textilinnovation gGmbH
Boennigheim, Germany

Yordan Kyosev
Institute of Textile Machinery and High Performance Material Technology (ITM)
Technische Universität Dresden
Dresden, Germany

Mona Maher
Department of Human Centered Design
Cornell University
Ithaca, New York

Mushfika Tasnim Mica
Wilson College of Textiles
North Carolina State University
Raleigh, North Carolina

P.Y. Mok
Institute of Textiles and Clothing
The Hong Kong Polytechnic University
Hunghom, Hong Kong

Simone Morlock
Hohenstein Institut für Textilinnovation gGmbH
Boennigheim, Germany

Abu Sadat Muhammad Sayem
Manchester Fashion Institute
Manchester Metropolitan University
Manchester, UK

Sin Ying NG
Institute of Textiles and Clothing
The Hong Kong Polytechnic University
Hunghom, Hong Kong

Evrim Buyukaslan Oosterom
Faculty of Applied Sciences, Textiles and Fashion Design
Istanbul Bilgi University
Istanbul, Turkey

Christian Pirch
Hohenstein Institut für Textilinnovation gGmbH
Boennigheim, Germany

Sanjukta Pookulangara
College of Merchandising, Hospitality and Tourism
University of North Texas
Denton, Texas

Kowshik Saha
Wilson College of Textiles
North Carolina State University
Raleigh, North Carolina

Annie Sautel
College of Merchandising, Hospitality and Tourism
University of North Texas
Denton, Texas

Yuyuan Shi
3D Weaving Innovation Centre, School of Design
University of Leeds Leeds, UK

Tatjana Spahiu
Textile and Fashion Department
Polytechnic University of Tirana
Tirana, Albania

Lindsey Waterton Taylor
3D Weaving Innovation Centre, School of Design
University of Leeds
Leeds, UK

Vanda Tomanova
Institute of Textile Machinery and High Performance Material Technology (ITM)
Technische Universität Dresden
Dresden, Germany

Part A

Introduction

1 Defining Digital Fashion and Tracking the Developments in Relevant Technologies

Abu Sadat Muhammad Sayem, Samit Chakraborty, S M Azizul Hoque, Kowshik Saha, Mushfika Tasnim Mica and Mominul Ahsan

CONTENTS

1.1 INTRODUCTION

Today's fashion supply chain is multi-continental and intertwined with diverse elements scattered in all over of the world. The best option, probably the only option, to make it truly transparent, sustainable, and effectively manageable is to digitalise it

DOI: 10.1201/9781003264958-2

seamlessly. The spread of digital fashion is not just limited to computer-aided design (CAD) and manufacturing (CAM), but rather runs throughout the fashion business, from product life-cycle management and developing new business models that promote sustainability to connecting virtual and augmenting reality with fashion to enhance consumers' experience through smart solutions (Sayem 2022, 139–141; Sayem et al. 2010, 45–53). It has emerged as a multidisciplinary field of knowledge that attracts overlapping interests from the academics, researchers and professionals coming from fashion design, business and technology, computer science, software engineering, animation and gaming, anthropometrics, supply chain management and industry 4.0, big data and artificial intelligence, and industrial sustainability.

1.2 DIGITAL FASHION AND ASSOCIATED THEMES

The adjective "digital" imparts different meanings to different nouns when it sits before them. Sometimes it refers to a fully non-physical and software entity like digital cinema, digital data, etc. Sometimes it implies physical devices and technology that capture, produce, store and/or display non-physical elements like digital camera, digital TV, digital clock, etc. Sometimes it indicates an approach or technique, although the device and the product involved are both physical, such as "digital printing". Likewise, the phrase "digital fashion" has different meanings in different contexts. Digitalisation is visibly present in different aspects of fashion design, manufacture, business, and product features. It is easy to perceive it through the use of digital technologies that are available for application in fashion industry. The technological innovations that are making digital fashion possible can be clustered under the four following themes: (1) digital design and e-prototyping, (2) digital business and promotion, (3) digital human and metaverse, and (4) phygital apparel and smart wearable technology. Figure 1.1 presents the technical areas associated with these four themes, which are further explored in the following sections.

1.3 DIGITAL FASHION DESIGN AND E-PROTOTYPING

Although, the era of CAD began with the advent of software systems named PRONTO in 1957 and Sketchpad in 1960, initial applications were limited to large aerospace and automotive companies due to the high price of early computers (Aouad et al. 2012). CAD systems became accessible to product designers once personal computers became available in the 1980s. In the same decade, AutoCAD and MiniCAD were introduced, and companies like Adobe and Corel were established. Since then, Adobe Illustrator and CorelDraw have gradually become popular digital tools for product design and graphic design. They are also extensively used by fashion designers. Different two-dimensional (2D) and three-dimensional (3D) CAD systems are used in the fashion industry. Table 1.1 lists the commonly used CAD systems in the fashion industry.

1.3.1 Two-Dimensional (2D) CAD for Illustration, Pattern Cutting and Marker Making

Fashion product development includes the development of design sketches and style outlines and weave, knit, embroidery, and print designs. For fashion and textile design,

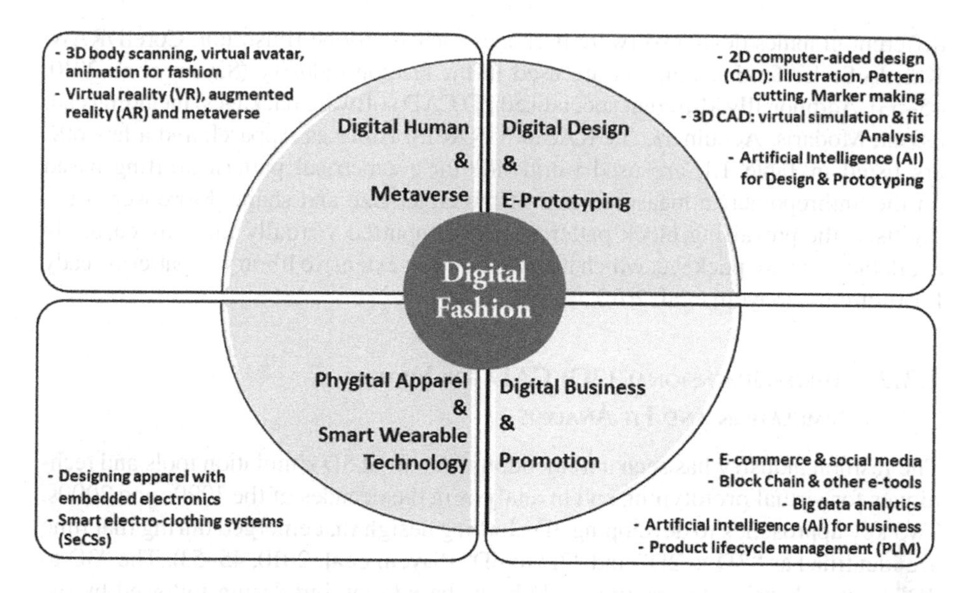

FIGURE 1.1 Major themes in digital fashion innovation.

Source: Adapted from Sayem (2022, 139–141).

TABLE 1.1
List of Available 2D and 3D CAD Systems for the Fashion Industry

CAD Systems					
2D CAD For Design and Illustration		2D CAD For Pattern Cutting and Marker Making		3D CAD For 3D Simulation and Visualisation	
System	Supplier	System	Supplier	System	Supplier
Corel Draw	*Corel*	PatternMaster	*Wild Ginger*	Style3D	*Linctex*
Illustrator	*Adobe*	Easy Pattern	*GT CAD*	CLO 3D	*Clo*
WeaveIt	*Canyon Art*	CreativeSTUDIO	*GeminiCAD*	Modaris	*Lectra*
Design Dobby/ Jacquard	*Textronic*	Gerber Accumark 2D	*Lectra*	TUKA 3D	*Tukatech*
Kaledo Style	*Lectra*	Moradis	*Lectra*	Vstitcher & Lotta	*Browzwear*
Colorsep, Pro-Weave, Knit	*Pointcarre*	PDS	*Optitex*	Gerber AccuMark 3D	*Lectra*
		TUKAcad	*Tukatech*		
Tex-Design	*Koppermann*	Telestia Creator	*SITAM-AB*	3D Vidya	*Assyst*
TUKAstudio	*TUKATECH*	PatternSmith	*Automatrix*	Modaris 3D	*Lectra*
Digital Fashion Pro	*Digital Fashion Pro*	PAD pattern design	*Pad system*	Tailornova	*Tailornova*
		Tailornova	*Tailornova*	3D Mapping	*Pointcarre*
		Pattern Xpert	StyleCAD	Design 3D	*Textronic*

different graphics design software packages such as Adobe Illustrator, CorelDRAW, Kaledo Style, Tex-Design, etc. are used in the fashion industry (Sayem et al. 2010, 45–53). Additionally, different specialised 2D CAD software packages, including cad. assyst, Modaris, Accumark, TUKAcad, GRAFIS, Audaces Apparel, and a few others listed in Table 1.1, are used worldwide for geometrical pattern drafting based on the anthropometric measurements of the target size and shape. Moreover, using digitiser, the prevailing block patterns can be inputted virtually into any currently available software package, which helps to build an extensive library of patterns ready for future use (Sayem et al. 2010, 45–53).

1.3.2 Three-Dimensional (3D) CAD for Virtual Simulation and Fit Analysis

The fashion industry has seen a tremendous growth of 3D simulation tools and techniques for virtual prototyping and fit analysis in the decades of the 1990s and 2000s. Two key approaches to developing 3D clothing design that emerged during that time are identified as "3D to 2D" and "2D to 3D" (Sayem et al. 2010, 45–53). The "3D to 2D" approach refers to creating a 3D body–based clothing design followed by the flattening of 3D shapes into 2D pattern pieces. On the other hand, "2D to 3D" is the more traditional approach where 2D digital pattern pieces are draped onto virtual mannequins or avatars. Sometimes these two approaches are combined to develop an advanced CAD system. A notable number of 3D CAD systems from companies such as Optitex, Gerber, CLO, Lectra, Browzwear, Tukatech, etc. are now commercially available for 3D designing or virtual prototyping (see Table 1.1).

1.3.3 Artificial Intelligence (AI) for Fashion Design and Prototyping

Artificial intelligence (AI) trains computer programs with data for advanced levels of application. Different AI methods (machine learning, decision support system, expert system, optimisation, and vision) have already been applied in textile and fashion design, fabric and apparel manufacturing, and wholesaling (Giri et al. 2019, 95376–95396; Chakraborty et al. 2021, 142–157; Chakraborty, Hoque et al. 2021, 1–43). The examples of using AI in fashion design and prototyping are rapidly increasing over time. In 2016, Google's collaboration with online fashion platform Zalando introduced "Project Muze". This project involved a predictive design engine that could make creative designs based on aesthetic parameters, including colour, style, and surface texture provided by the fashion experts. Cross & Freckle, a New York–based fashion upstart, made T-shirts designed by AI. An entire collection of Balenciaga was recently developed by Robbie Barrat using deep neural network, where he used thousands of photos from the brand's look books and previous fashion shows (Monie 2021). Unspun Inc., a California-based start-up, collaborated with H&M to make custom fit jeans that are unique to each consumer's body (Roberts-Islam 2020). Researchers are also exploring new ideas and opportunities to integrate AI in fashion design and prototyping. Jeon et al. (2021) proposed AI model "FashionQ" based on three cognitive processes, namely extending, constraining, and blending. It is a creative support tool that facilitates ideation in fashion design based on divergent and convergent thinking. Kato et al. (2018) developed a deep neural network–based model named DeepWear that can generate patterns based on the clothing features learnt from a particular brand's clothing images. It is expected to see a rapid expansion of AI in the field of fashion design and virtual prototyping in the near future. AI is helpful in clothing recommendation systems, prediction of material quality, creating personalised shopping experiences through virtual and augmented reality, defect detection, sales forecasting, and trend analysis. Also, AI is useful in real-time defect detection, making the manufacturing process more sustainable by reducing wastage (Chakraborty et al. 2021, 142–157).

1.4 DIGITAL FASHION BUSINESS AND PROMOTION

Digital fashion business has evolved at an accelerated rate in the last few years to accommodate consumers' expectations and business dynamics. One notable transition that drastically impacted the market is e-commerce. The fashion industry is the fourth largest sector globally and worth more than 2.4 trillion dollars. Traditional brick-and-mortar stores are being replaced by virtual platforms to serve this big market in this digital era. Consumers are now utilising online platforms to make purchase decisions through exploring, comparing, reading reviews, virtual payment, and home delivery. Hence, fashion brands are adopting virtual tools for advertising their products and offers, offering easy design and fitting exploration through virtual reality (VR) and augmented reality (AR), accepting virtual payments, and offering easy delivery and return policies (Boardman et al. 2020, 155–172). Some companies are implementing an omnichannel approach that takes advantage of both offline and online platforms to offer a smooth shopping experience (Ryu et al. 2019, 74–77). All these adaptations have made the fashion (apparel, accessories, and footwear) industry the number one e-commerce sector, with a 759.5 billion dollar market value in 2021, which is expected to reach one trillion dollars by 2025 (Orendorff 2021). Therefore, digital activities are very crucial for the success of fashion brands.

1.4.1 E-Commerce and Social Media in Fashion Business

E-commerce in the fashion business utilises virtual platforms for business activities, that is, marketing, communication, and transaction. It has created opportunities for fashion brands to better understand consumers' expectations, offer a wide range of products with easy customisation, reach beyond geographic boundaries, communicate quickly, and complete transactions easily (Kalbaska and Cantoni 2019, 125–135). Social media marketing has become a popular choice to reach the target market. According to a statistics of 2021, 56.8 percent of the world's population (4.48 billion people) use different social media platforms, among which Facebook alone had 2.9 billion monthly users. This statistic highlights the potential of social media marketing. Facebook has been the most preferable platform, while fashion brands also use YouTube, Instagram, and Twitter (Mazzucchelli et al. 2021, 1107–1144). According to *Entrepreneur* magazine, almost all Fortune 500 companies (97%) are actively present in social media platforms (Porteous 2021). Marketing on these platforms involves the advertisement of products, offers and relevant information to the virtual community through an interpersonal network, virtual forums, and interactive communication by means of words, pictures, audio, and videos. Reviews and discussions help brands observe consumers' perceptions on offered products and promotional activities (Nash 2018, 82–103). Moreover, social media analytics assists in trend analysis, fashion forecasting, and monitoring evaluation (Chakraborty et al. 2020, 376–386). All these analyses can guide brands to strategise appropriately. As a result, fashion brands are dedicating a team for social media activities.

1.4.2 Big Data Analytics

Big data has been a significant tool in the age of fashion digitalisation. It can be defined as the constant collection of an enormous amount of data, which are analysed to extract meaningful insight. Gathering data on the current market, consumer preferences, and upcoming trends helps fashion brands to understand the market better and be proactive with their approach. Hence, big brands like Zara, ASOS, H&M, Macy's, and Gap are constantly taking advantage of big data through market analysis, trend forecasting, design development, and advertising (Silva et al. 2020, 21–27). It has made the prediction of future trends more robust (DuBreuil and Lu 2020, 68–77). Big data can easily track their most trendy item, observe consumer feedback, and track consumer behaviour. Therefore, fashion brands are investing more and more in data analytics and making decisions based on that.

1.4.3 Product Lifecycle Management

Product lifecycle management (PLM) records, tracks, manages, and shares all the information of a product's value chain encompassing its whole lifecycle (Terzi et al. 2010, 360–389). Digitalisation has brought great flexibility to PLM. With digitalisation, any party within a PLM system can update and share real-time information with all parts of a value chain, which is helpful in designing, planning, scheduling, and costing. Fashion brands are utilising this technology to ensure proper communication regarding the product's expectations, requirements, developments, and progress among different departments. As a result, proper coordination can be achieved that results in quality work and cost savings. Some of the available digitalised PLM for fashion brands include YuniquePLM by Gerber technology, BlueCherry by CGS Inc., InforPLM by Infor, CentricPLM by Centric, LectraPLM by Lectra, etc. These PLM platforms are becoming integral components in fashion production.

1.4.4 AI and Blockchain Technology

In addition to trend forecasting, AI can assist fashion companies in making better decisions and aid online shoppers during their browsing and product searching activities. Another advanced technology that is facilitating the fashion industry is blockchain technology. Blockchain technology can create a secure network among different tiers of the fashion supply chain and track the transections among them (Agrawal et al. 2021, 1–12). One example of such a platform is TextileGenesis™.

They are enabling transparency in the fashion supply chain utilising their patent-pending blockchain technology named Fibercoins™ (TextileGenesis™, n.d.). These technologies are imparting higher efficiency and sustainability towards the fashion industry.

1.5 DIGITAL HUMAN AND METAVERSE

Creating a virtual environment using 3D CAD is not a new concept anymore, as such technology solutions have been in existence for several years now. Participating in games and events in the virtual environment with the help of a joystick and relevant gadgets also has been known to us for a notable period of time. But appearing and existing in the virtual environment as a virtual avatar and roaming around within the virtual world is the concept of *metaverse*. This has already started to happen, and it has opened a new horizon of digital fashion.

1.5.1 3D Body Scanning, Virtual Avatars, and Fashion Animation

3D body scanning devices that initially arrived as a solution for contactless body measurement offer an easy, probably the best, way of capturing 3D geometry and the architecture of the human body with fine details for potential integration into the virtual world. A wide range of systems—with static or dynamic scan head, single or multiple scan head, handheld or with turntable, portable or not portable—are available from different suppliers, including TC2, Cyberware, Size stream, Telmat, Styku, Intelfit, Virtonic, Arctec, and Cad Modelling. At the same time, body scanning devices with real-time motion capturing capability, such as Move4D, are also available on the market. These provide an easy and quick way of producing human avatars with a near-to-reality condition.

1.5.2 Virtual Reality (VR), Augmented Reality (AR), and Metaverse

VR and AR are commonly clustered within the umbrella term "Extended Reality (XR)". With the help of a headset, VR can connect and place us in a computer-generated world which can be virtually explored, whereas AR makes use of digital images and layers them on the real world with the help of a visor or smartphone. VR technology has been adopted by fashion retailers to promote virtual runway shows to store visitors and virtual display of products. VR elevates the customer experience by facilitating virtual try-on clothing that eliminates taking off any of their garments, as well as changes the colours and sizes of garments (Hwangbo et al. 2017, 1–17). Available AR applications range from virtual closets to digital clothing and include virtual showrooms, virtual fitting rooms, mobile scanning, face tracking, digital clothing, and interactive and digital display. Tech providers like Zugara, 3DLook, Nettelo, and Atlatl offer AR assisted try-on applications and have been massively adopted in the retail sectors both publicly and privately (Javornik 2016, 252–261).

1.6 PHYGITAL APPAREL AND SMART WEARABLE TECHNOLOGY

Phygital apparel can be defined as an interface between the body and the digital system, also known as smart electro-clothing systems (SeCSs). Thanks to the ongoing miniaturisation of electronic components and evolving new technologies of Internet of Things (IoT), different SeCSs are being developed for health, fitness, sports, and social benefits (Sayem et al. 2020, 1–23).

1.6.1 Designing Phygital Apparel with Embedded Electronics

Designing phygital apparel with embedded electronics requires the knowledge of textile fibre, yarn, and fabric constructions, as well as the appropriate use of sensors based on the application at the correct position of the wearer (Sayem et al. 2020, 1–23). Figure 1.2 presents a framework for designing SeCSs proposed by Ahsan et al.

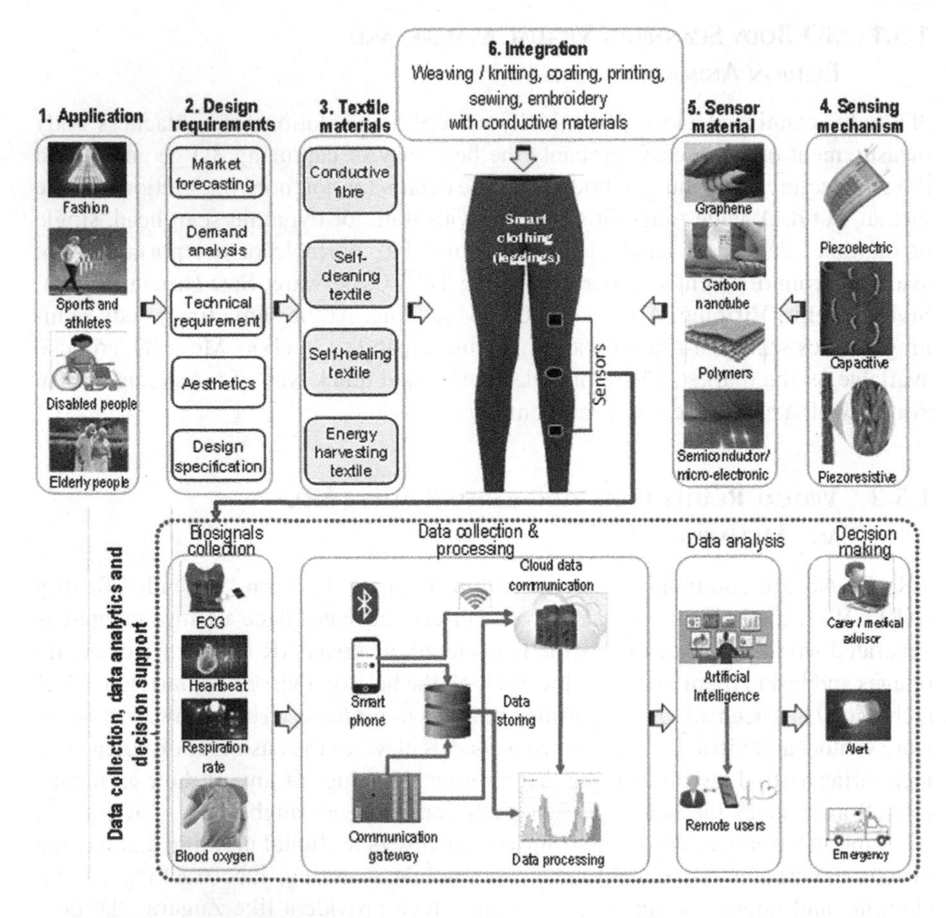

FIGURE 1.2 A conceptual framework for developing smart clothing.

Source: Ahsan et al. (2022, 113–145), Creative Common Licence.

(2022, 113–145). Textile structures, such as those woven or knitted from conductive yarns, and conductive print-inks, including graphene-based print inks, assist in integrating lightweight sensors onto textiles. Micro-electromechanical systems (MEMS) are defined as miniature integrated devices or systems that integrate mechanical and electrical components. Typically, MEMS are mechanical microstructures like microsensors and micro actuators that are integrated onto the same silicon chip (Bogue 2013, 300–304). Several sensors, such as textile and non-textile, can be used for phygital apparel development. Textile fabrics serve as the core substrate for integrating other subsystems in and on them to build SeCSs. Any sensor must be highly and selectively sensitive to biopotential (e.g., ECG, EMG, and EEG) or other specific indicators. The transporting power and bio-signal interconnect the sensor point and the data processing unit (electronic board). Metallic wires and components have traditionally been used as interconnectors for electrical products, but textile-based interconnects are becoming more popular for SeCSs to meet the consumer use criteria.

1.6.2 Smart Electro-Clothing Systems (SeCSs)

The system architecture of a SeCS generally includes eight working subsystems and two supporting subsystems (see Figure 1.3). The standard working subsystems included in a SeCS are as follows: (1) control subsystem, (2) sensing subsystem, (3) actuator subsystem, (4) communication subsystem, (5) location subsystem, (6) power subsystem, (7) storage subsystem, and (8) display subsystem. The supporting subsystems include interconnection and software subsystems. Although most of the constituent components of these subsystems, such as the control subsystem, certain types of sensing and actuator subsystems, location subsystem, power subsystem, storage subsystem, and display subsystem, are electronics and non-textile materials, they (except the display subsystem) can be accumulated within an electronic board in miniaturised forms to finally connect to textile components (Sayem et al. 2020, 1–23).

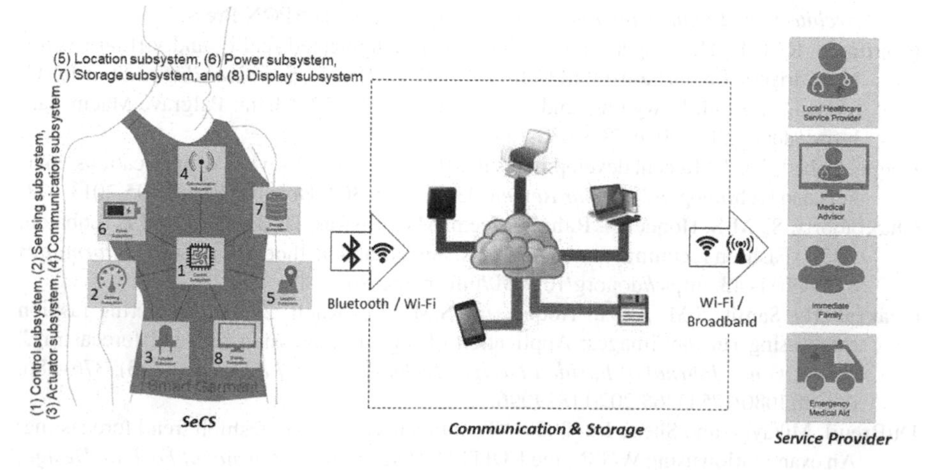

FIGURE 1.3 Typical SeCS architecture.

Source: Sayem et al. (2020, 1–23), Creative Common Licence.

SeCSs have applications in several areas, including health, sports, and social activities (Sayem et al. 2020, 1–23). Medical disorders can be detected and monitored using textile-based devices that can sense bio-signals such as ECG, body temperature, and so on. "SeCSs for sports" are systems that their providers push for sports applications, such as monitoring players' and athletes' physical conditions and performance, as well as assisting players/athletes and their coaches in training and coaching. "SeCSs for fitness" are systems that help typical customers with daily fitness activities, such as walking, jogging, running, and performing yoga. Also proposed for social applications are different technologies that assist users in their social activities (Sayem et al. 2020, 1–23).

1.7 CONCLUSION

Digital fashion technologies are continuously evolving and disrupting the fashion industry. The recent Covid-19 pandemic accelerated the adoption of digital technologies in different industries. However, this adoption is not taking place equally and at an even pace throughout the fashion supply chain, which spreads over multiple continents. At the same, the need for new skill sets and training programmes is seriously felt within the industry for implementing such technologies.

REFERENCES

Agrawal, T.K., V. Kumar, R. Pal, L. Wang, and Y. Chen. 2021. "Blockchain-based framework for supply chain traceability: A case example of textile and clothing industry." *Computers & Industrial Engineering* 154 (107130): 1–12. https://doi.org/10.1016/j.cie.2021.107130.

Ahsan, M., S.H. Teay, A.S.M. Sayem, and A. Albarbar. 2022. "Smart clothing framework for health monitoring applications." *Signals* 3: 113–145.

Aouad, Ghassan, Song Wu, Angela Lee, and Timothy Onyenobi. 2012. "Chapter 1, introduction to CAD for the AEC/FM industry." In *Computer Aided Design Guide for Architecture, Engineering and Construction*. New York: SPON Press.

Boardman, R., C.E. Henninger, and A. Zhu. 2020. "Augmented reality and virtual reality: New drivers for fashion retail?" In *Technology-Driven Sustainability*, edited by G. Vignali, L. Reid, D. Ryding, and C. Henninger, 155–172. Cham: Palgrave Macmillan. https://doi.org/10.1007/978-3-030-15483-7_9

Bogue, Robert. 2013. "Recent developments in MEMS sensors: A review of applications, markets and technologies." *Sensor Review* 33 (4): 300–304. Doi: 10.1108/SR-05-2013-678.

Chakraborty, S., M.S. Hoque, N. Rahman Jeem, M.C. Biswas, D. Bardhan, and E. Lobaton. 2021. "Fashion recommendation systems, models and methods: A review." *Informatics* 8, 3 (49): 1–43. https://doi.org/10.3390/informatics8030049.

Chakraborty, Samit, S.M. Azizul Hoque, and S.M. Fijul Kabir. 2020. "Predicting fashion trend using runway images: Application of logistic regression in trend forecasting." *International Journal of Fashion Design, Technology and Education* 13 (3): 376–386. doi: 10.1080/17543266.2020.1829096.

DuBreuil, Mikayla, and Sheng Lu. 2020. "Traditional vs. big-data fashion trend forecasting: An examination using WGSN and EDITED." *International Journal of Fashion Design, Technology and Education* 13 (1): 68–77. doi: 10.1080/17543266.2020.1732482.

Giri, C., S. Jain, X. Zeng, and P. Bruniaux. 2019. "A detailed review of artificial intelligence applied in the fashion and apparel industry." *IEEE Access* 7: 95376–95396. doi: 10.1109/ACCESS.2019.2928979.

Hwangbo, H., S.Y. Kim, and K.J. Cha. 2017. "Use of the smart store for persuasive marketing and immersive customer experiences: A case study of Korean apparel enterprise." *Mobile Information Systems* 2 (17): 1–17.

Javornik, Ana. 2016. "Augmented reality: Research agenda for studying the impact of its media characteristics on consumer behaviour." *Journal of Retailing and Consumer Services* 30: 252–261.

Jeon, Youngseung, Seungwan Jin, Patrick C. Shih, and Kyungsik Han. 2021. "FashionQ: An AI-driven creativity support tool for facilitating ideation in fashion design." Proceedings of the 2021 CHI Conference on Human Factors in Computing Systems (CHI '21). Article 576, 1–18. Association for Computing Machinery, New York, NY, USA. https://doi.org/10.1145/3411764.3445093

Kalbaska, N., and L. Cantoni. 2019. "Digital fashion competences: Market practices and needs." In *Business Models and ICT Technologies for the Fashion Supply Chain*, edited by R. Rinaldi and R. Bandinelli, 125–135. Cham: Springer International Publishing. https://doi.org/10.1007/978-3-319-98038-6_10

Kato, Natsumi, Hiroyuki Osone, Daitetsu Sato, Naoya Muramatsu, and Yoichi Ochiai. 2018. "DeepWear: A case study of collaborative design between human and artificial intelligence." Proceedings of the Twelfth International Conference on Tangible, Embedded, and Embodied Interaction (TEI '18), 529–536. Association for Computing Machinery, New York, NY, USA. https://doi.org/10.1145/3173225.3173302

Mazzucchelli, A., R. Chierici, A. Di Gregorio, and C. Chiacchierini. 2021. "Is Facebook an effective tool to access foreign markets? Evidence from international export performance of fashion firms." *Journal of Management and Governance* 25 (4): 1107–1144. https://doi.org/10.1007/s10997-021-09572-y

Monie, Karine. 2021. "Your AI-generated clothes are trending." *WIRED*. Accessed October 17, 2022. https://wired.me/culture/design/your-ai-generated-clothes-are-trending/.

Nash, J. 2018. "Exploring how social media platforms influence fashion consumer decisions in the UK retail sector." *Journal of Fashion Marketing and Management: An International Journal* 23 (1): 82–103. https://doi.org/10.1108/JFMM-01-2018-0012.

Orendorff, Aaron. 2021. "10 trends styling 2022's ecommerce fashion industry: Growth + data in online apparel & accessories market." *Common Thread*. Accessed October 17, 2022. https://commonthreadco.com/blogs/coachs-corner/fashion-ecommerce-industry-trends.

Porteous, Chris. 2021. "97% of fortune 500 companies rely on social media: Here's how you should use it for maximum impact." *Entrepreneur*. Accessed October 16, 2022. www.entrepreneur.com/article/366240.

Roberts-Islam, B. 2020. "H&M Group experiments with made to order 'fast fashion' to tackle overstock." Accessed October 17, 2022. www.forbes.com/sites/brookerobertsislam/2020/11/26/hm-group-experiments-with-made-to-order-fast-fashion-to-tackle-overstock/?sh=285a125523fa.

Ryu, M.H., Y. Cho, and D. Lee. 2019. "Should small-scale online retailers diversify distribution channels into offline channels? Focused on the clothing and fashion industry." *Journal of Retailing and Consumer Services* 47: 74–77. https://doi.org/10.1016/j.jretconser.2018.09.014.

Sayem, Abu Sadat Muhammad. 2022. "Digital fashion innovations for the real world and metaverse." *International Journal of Fashion Design, Technology and Education* 15 (2): 139–141. doi: 10.1080/17543266.2022.2071139.

Sayem, Abu Sadat Muhammad, Siew Hon Teay, Hasan Shahariar, Paula Luise Fink, and Alhussein Albarbar. 2020. "Review on smart electro-clothing systems (SeCSs)." *Sensors* 20 (3): 587, 1–23. https://doi.org/10.3390/s20030587.

Sayem, Abu Sadat Muhammad, Richard Kennon, and Nick Clarke. 2010. "3D CAD systems for the clothing industry." *International Journal of Fashion Design, Technology and Education* 3 (2): 45–53. doi: 10.1080/17543261003689888.

Silva, E.S., H. Hassani and D.Ø. Madsen. 2020. "Big data in fashion: Transforming the retail sector." *Journal of Business Strategy* 41 (4): 21–27. https://doi.org/10.1108/JBS-04-2019-0062.

Terzi, S., A. Bouras, D. Dutta, M. Garetti, and D. Kiritsis. 2010. "Product lifecycle management: From its history to its new role." *International Journal of Product Lifecycle Management* 4 (4): 360–389. https://doi.org/10.1504/IJPLM.2010.036489.

TextileGenesis™. n.d. "White paper: The business case for traceability!" Accessed October 17, 2022. https://textilegenesis.com/.

Part B

Digital Design and E-Prototyping

2 Clothing Fit Evaluation
From Physical to Virtual

Abu Sadat Muhammad Sayem

CONTENTS

DOI: 10.1201/9781003264958-4

2.1 INTRODUCTION

Clothing fit is undoubtedly a vital issue for consumers, manufacturers, and retailers. Consumers often describe it as a synonym of comfort and quality. There have been a notable number of published reports highlighting the prevailing problem of ill-fitting clothing, consumers' dissatisfaction over clothing fit, and/or their difficulties in finding clothing which fits correctly. According to an industry report published in 2000, half of all American consumers struggled to find clothing with satisfactory fit in the marketplace (DesMarteau 2000). This problem has never declined as the *Chicago Tribune* reported in 2007 that about 84 percent of female consumers claimed not to be able to find clothing that fits properly, and poor or inconsistent fit accounted for a huge monetary loss in women's apparel sales (Giovis 2007). In the case of mail or catalogue orders, a return rate of 30 percent and 50 percent has been reported in 1993 (Abend 1993, 74–75) and in 2000 (DesMarteau 2000), respectively, due to fit problems. In the USA, 23 percent of all apparel purchases are returned, and 46 percent of the customers who returned their purchased clothing mentioned lack of fit as the major reason for returning (Cilley 2016).

Fit evaluation is an indispensable part of the clothing development and manufacturing process. Considering the extent of the fit-related problem within the apparel industry, this has been an area of continuous interest to industry professionals and researchers. This chapter reviews and summarises the prevailing concepts of clothing fit and its interrelated factors, along with consumers' fit preference issues and fit evaluation processes, both physical and virtual.

2.2 DEFINITION AND COMPONENTS OF CLOTHING FIT

Divergent definitions of clothing fit prevail within the industry (Yu 2004a). According to an early definition presented in Yu (2004a), fit is directly related to human anatomy, and most of the fit problems arise from the bulges of the body (Cain 1950). However, natural body bulges cannot be blamed for fit problems, rather the ability of a garment to comply with those bulges is responsible for a good fit or a misfit. Well-fitted clothing conforms to the wearer's body, includes adequate ease of movement, shows no undesirable wrinkles, and has been made in such a way that it appears to be part of the wearer (Yu 2004a). Clothing fit can be characterised by the presence of a neat and smooth appearance and the assurance of maximum comfort and mobility for the wearer (Shen and Huck 1993). Clothing fit can be considered as a state of mind as well as a state of physical being, and a wearer feels physically and psychologically uncomfortable when wearing clothes that do not fit (Rasband and Liechty 1994). Chen (2007), similar to Cain (1950), identified fit as the relationship between the size and contour of the garments and those of the human body. The physical and psychological attributes linked with comfort and satisfaction are involved in the assessment of clothing fit (Ashdown and O'Connell 2006). According to Kincade (2008), 'fit is a measure of the conformance of the product to the body, with consideration for ease of style and ease of movement'.

It is apparent from the aforementioned definitions that clothing fit is composed of three integral components: anthropometry, comfort, and clothing manufacture (Ashdown and O'Connell 2006; Chen 2007; Shen and Huck 1993). Each of these three components is related to several factors, some of which are inter-related (see Figure 2.1).

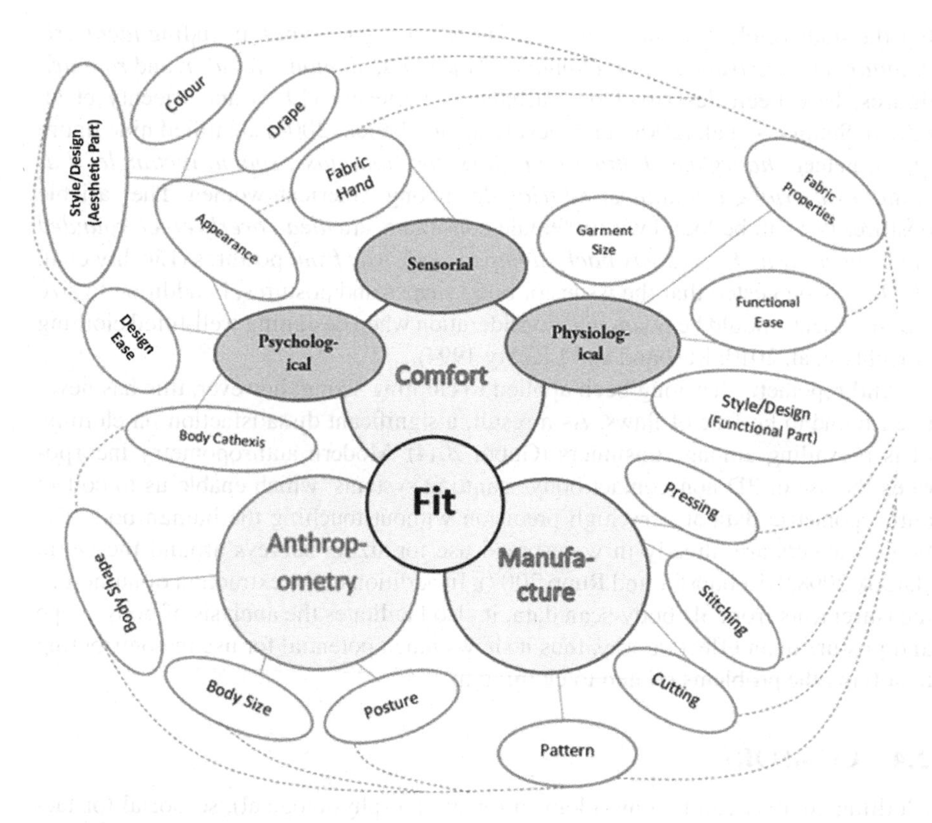

FIGURE 2.1 Factors related to clothing fit.

2.3 ANTHROPOMETRY

Anthropometry, a branch of ergonomics, refers to the scientific measurement and collection of data about physical characteristics of the human body. Accurate acquisition of anthropometric information and its efficient interpretation into pattern pieces forms the foundation of clothing fit. It is apparent from anthropometric analysis that differential body sizes, shapes, and postures exist within the human population. For example, a survey conducted by German DOV-Verband in 1993 identified three different hip types, namely slim, standard, and broad hips, among the adult female population, distributed in percentages of 36, 42, and 22, respectively (Bougourd 2007). Different body sizes within the same age group and different body shapes within the same size group are quite common irrespective of ethnicity. Again, human figures can be vertically and/or horizontally balanced or unbalanced (Liechty et al. 2010). An early classification of the male body, which depicted body types as *endomorph*, *mesomorph*, and *ectomorph*, was done based on the presence or absence of body fat (Sheldon 1954). This was later modified and adapted for female figures by Johnson (1990). However, this classification has not been adopted within the fashion industry, which prefers to classify body types based on relative width or contours of the body.

For the study of clothing fit, eight different female figure types, including *ideal, triangular, inverted triangular, rectangular, hourglass, diamond, tubular,* and *rounded* figures, have been described by Rasband and Liechty (1994) and Liechty et al. (2010). Simmons et al. (2004) and Devarajan and Istook (2004) identified nine figure types, namely *hourglass, bottom hourglass, top hourglass, spoon, rectangle, diamond, oval, triangle,* and *inverted triangle,* among American women. The variable postures that can be found within female consumers are *ideal, overly erect, rounded upper back, sway back, sway back, slumped,* and *sway front* postures (Liechty et al. 2010). It is suggested that the issues of body shapes and postures, in addition to size measurement, should be taken into consideration when designing well-fitted clothing (Liechty et al. 2010; Rasband and Liechty 1994).

Anthropometry has long been applied to clothing sizing; however, this has never been found to be free of flaws. As a result, a significant dissatisfaction on clothing fit is prevailing among consumers (Gupta 2014). Modern anthropometry incorporates the use of 3D non-contact body-scanning systems, which enable us to collect anthropometric data at very high precision without touching the human body. 3D body scanners are already in widespread use for sizing surveys around the world (Istook 2008; Kirchdörfer and Rupp 2007). In addition to the extraction of numerous measurements from 3D body-scan data, it also facilitates the analysis of body shape and posture in an effective way, thus it shows much potential for use in contributing to solving the problems related to clothing fit.

2.4 COMFORT

Clothing comfort can be physiological (or thermo-physiological), sensorial (or tactile), and psychological, and it is a key component of clothing fit when the satisfaction of a wearer of clothing is concerned.

2.4.1 Physiological Comfort

Helping the human body maintain its thermal balance and thermo-physiological comfort is a fundamental function of clothing—especially for functional or performance clothing (Li 2001; Ho et al. 2011). Physiological clothing comfort can be influenced by the selection of the appropriate garment size, style/design, fabric properties, and functional ease incorporated in the patterns to allow for the wearer's activities (body movements). These factors are further detailed in the following sub-sections.

2.4.1.1 Garment Size

Garment size is determined based on body size, which is the subject of anthropometry. Selection of the wrong garment size means excessive differences in vertical, horizontal, and circumferential measurements, either too small or too large, between the wearer's body and the garment. This influences the amount of air gap and trapped still air in between the body and the garment, ventilation efficiency, the thickness of the microclimate around the human body (Das and Alagirusamy 2010), and ultimately the thermal insulation value of the garment.

2.4.1.2 Style/Design (Functional Part)

Garment design, in terms of style and design details, plays a crucial role in ensuring thermal comfort of clothing (Ho et al. 2011). The size of openings at different places of a garment, such as the neck, waist, wrists, and ankles, can influence the thermal insulation and moisture vapour resistance of garments during windy conditions and body movement. In general, a garment with large openings is expected to have reduced thermal insulation and moisture vapour resistance. Conversely, with small and closed openings, even a loose-fitting garment can have much greater thermal insulation than a tight-fitting garment of similar style (Ho et al. 2011; McCullough et al. 1983). Without changing the quality of fabrics, a designer has ample options to technically manipulate the thermal insulation, moisture vapour resistance, and convective heat transfer in clothing by efficiently manipulating style and design details (Ho et al. 2011).

2.4.1.3 Fabric Properties

Thermal insulation and moisture vapour resistance are two major properties of fabrics that influence the thermal comfort of clothing. These two depend on fabric characteristics and construction, such as structure (woven, knitted, or non-woven) and thickness, which are very closely influenced by fibre and yarn types and their properties. Even though fabric properties influence the clothing fit by playing a role in wear comfort, there is no objective means of considering them in the pattern design process. Thus, the clothing fit is highly dependent on the expertise of the pattern technician and the efficiency of the production system.

2.4.1.4 Functional Ease

Ease is the additional measurement added to pattern pieces. There are two types of ease—functional (wearing) ease and design (styling) ease (Kincade 2008). Functional ease is necessary to allow the body within a garment to move freely without any constraint. Commonly it is added with a pattern of the garment parts that cover to expandable body parts, for example, chest expansion while breathing in and out. How much functional ease can be added with a pattern is determined by several factors, such as the type of garment, the type of fabric, and the gender and activities of the wearer. On the other hand, design ease is added to pattern pieces to reflect the garment silhouette imagined by fashion designers. Ease can even be negative in certain cases for close-fitting garments if they are made of stretch fabric containing elastane fibres. Usually design sketches are not done in full-scale size, therefore there is no question of calculating ease at the design stage. Designers only give a subjective outline of the design ease, but this does not include any functional ease. It is the job of the pattern technicians to calculate the right amount of ease to add with the pattern edges with consideration for the designer's concept, the fabric properties, and the consumer's requirements. However, there is no rule except relying on experience and common practices to determine the exact amount of ease for incorporation in pattern pieces (Petrova and Ashdown 2008). Sometimes clinical studies provide clues about functional ease requirements in clothing. For example, Moll and Wright (1972) studied the normal chest expansion during breathing in men and women of different ages.

Their study reveals that chest expansion during breathing varied by age and gender. It increases as we grow older, up to 34 years of age. After the age of 35, chest expansion during breathing follows a decreasing trend as we grow older. Later in 2002, Mckinnon and Istook (2002) noticed that the changes in circumferential measurements of the chest/bust, neck, and waist varied during inhalation and exhalation while investigating the effect of respiration on dimensional changes of participants using body-scanned data. Their study also provided clues about calculating functional ease in pattern pieces to facilitate comfortable breathing of wearers. The effect of breathing was also considered by Koblyakova (1980) to calculate minimum ease values in clothing to ensure comfortable movement for upper-body garments.

Similar to breathing, any intended or unintended body movement can cause change in shape and skin dimensions (Chi and Kennon 2006; Kirk and Ibrahim 1966). Arm movements in different directions cause significant dimensional changes of skin around the armhole, and this is an important considerable factor while calculating ease for garment sleeves, especially for functional garments (Chi and Kennon 2006). There are also other determining factors for calculating ease allowance. The mechanical properties of fabrics, especially the extension under tension, are also deciding factors for calculating the amount of ease in different parts of clothing (Ziegert and Keil 1988). In the case of close-fitting garments, stretchability can directly impact on wearers' comfort and freedom of movements, so it is also taken into account while determining functional pattern ease.

2.4.2 Sensorial Comfort

Sensorial comfort describes how a fabric or garment feels when it is worn, and fabric handle attributes (commonly known as fabric hand) influence the magnitude of comfort sensation experienced by a wearer (Das and Alagirusamy 2011). The mechanical and surface properties which can influence fabric hand and sensorial comfort of garments are uniaxial and biaxial tensions, shear under tension, bending, lateral compression, longitudinal compression and buckling, and surface roughness and friction (Postle 1983). Kirk and Ibrahim (1966) investigated the relationship between fabric extensibility and anthropometric kinematics (the body dynamics) in a garment and found that fabric stretch and friction played a role in consumers' comfort from clothing. Through the development of an objective measurement system of fabric properties in the 1970s, known as the Kawabata Evaluation System (KES), the quantitative relationship between fabric properties, tailorability, and mechanical comfort of garments was established (Kawabata and Niwa 1989; Kawabata et al. 2002). Niwa et al. (1998) related the fabric mechanical properties with the design of optimum silhouettes of ladies' garments by developing an *objective discriminant equation* which utilised the values of fabric properties measured by the Kawabata System. Their equation provided a guideline for fabric selection while designing a silhouette of ladies' garments.

2.4.3 Psychological Comfort

Psychological clothing comfort is guided by wearers' emotions and affection (Das and Alagirusamy 2010). Factors that are related to psychological comfort include clothing appearance (clothing aesthetics), amount of design ease that ultimately influences the clothing's appearance, and body cathexis. Clothing appearance and style is a subject of fashion trends, which is one of the primary criteria of clothing selection by a consumer. Clothing appearance reflects the aesthetics of clothing, which is a combination of style, colour, and drape. Design ease contributes to the silhouette of garments, which is also a subject of fashion trends. LaBat and Delong (1990) found that body cathexis, the positive and negative feelings towards one's body, is an impacting factor on the consumer's satisfaction with clothing fit.

2.5 MANUFACTURE

The major issues in manufacture that particularly affect clothing fit are pattern creation, fabric cutting, stitching, and pressing. The traditional pattern drafting technique depends on the size tables, which include only the measurements of certain positions on the human body, but does not represent the 3D shape and posture. This is a major deficiency of the pattern drafting technique, which makes it heuristic in nature. At the same time, any error in any of the sub-processes of manufacturing, such as cutting, stitching, and pressing, can result in the poor fit of garments. The deficiencies in the manufacture of clothing include fit model selection, pattern development, grading, grain distortion, and improper construction, which may cause a fit problem (Ashdown and O'Connell 2006).

2.6 INTERCONNECTED FACTORS

Some factors of the three components of clothing comfort are interconnected to each other, as can be seen in Figure 2.1. Pattern cutting is a part of clothing manufacture and is dependent on body size, shape, and posture, which belong to anthropometry, although mass production of off-the-peg clothing mainly utilises the average body size of a target population. Pattern creation, on the other hand, is a function of fabric properties, which a pattern technician always needs to keep in mind while drafting a pattern piece for any particular garment. Fabric properties being an influential factor of physiological comfort significantly determine its handle characteristics, which are factors of sensorial comfort, as well as clothing appearance that fall into the psychological component of clothing comfort. Body cathexis is a psychological interpretation of one's perception about his/her own body, which we study as a part of anthropometry.

2.7 GARMENT-FIT CLASSIFICATION

There are different fit-types for apparel items that range from slim and form-fitting to oversized. Based on the closeness of fit, Liechty et al. (2010) classified garment styles as *very fitted*, *slightly fitted*, *slightly loose*, *loose*, or a combination—*semi-fitted* or *partially fitted*. According to Bubonia (2014), different fit types of tops and dresses for American women are *form fitting/fitted*, *semi-fitted/relaxed*, *oversized/curvy*, *natural*, and *loose fitting/boxy*; and of bottoms they are *slim*, *natural*, *relaxed*, and *oversized*. Chattaraman et al. (2013) commissioned a field study in the American fashion industry and identified four or five industry-relevant fit classifications of different products, such as jeans (*skinny*, *slim*, *regular*, *relaxed*, and *loose*), khakis (*slim*, *classic*, *straight*, and *relaxed*), dress shirts (*fitted*, *slim*, *modern*, and *regular*), and polos (*fitted*, *slim*, *regular*, and *loose*). It is also evident from the websites of European retailers such as M&S, H&M, C&A, and Debenhams, etc. that a nearly similar classification of fit of different clothing items prevails on the market.

2.8 CONSUMERS' FIT PREFERENCE

Clothing fit preference is a very subjective matter. It varies from person to person (Alexander et al. 2005, 52–64). Physical, demographic, and psychological factors significantly influence the fit preferences of both male and female consumers (Alexander et al. 2005; Chattaraman and Rudd 2006; Pisut and Connell 2007; Simmons and Istook 2003). This is further elaborated in the following sub-sections.

2.8.1 Influence of Physical Factor (Body Size and Shape)

It has been found that body size has a positive relationship with choice of loose fit and body coverage by clothing. A study on American male consumers (19 to 66 years of age) reveals that consumers with higher BMI (body mass index), that is, bigger body size, prefer looser fit in garments and more body coverage by garments; an increase in age and body dissatisfaction also positively influences the preference of looser fits and more body coverage by garments (Chattaraman et al. 2013). Another study by Chattaraman and Rudd (2006) on American female undergraduate students (ranging in age from 18 to 56 years and mostly Caucasians, but also includes a small percentage of African American, Hispanic, and Asian American) found that female consumers with larger upper body size preferred tops with looser silhouettes, and those with a larger lower body size preferred a looser fit and higher waist level for bottoms.

In addition to size, body shape also influences consumers' choice on clothing fit. According to the findings of a study on American adult women (aged 19–54), the women with self-designated rectangular or pear-shaped bodies prefer loose-fitted garments, whereas hourglass or inverted triangular shaped bodies more likely prefer fitted garments (Pisut and Connell 2007). Yoo (2003) found that American working women (age 22 years and older) with a diamond shape body preferred a loosely fitted silhouette, whereas women with other body types mostly preferred fitted and semi-fitted jacket silhouettes.

2.8.2 Demographic Factors (Age)

Manuel et al. (2010), through a study on African American women, found that consumers in their 20s preferred a fitted jacket, whereas those in their 30s preferred a loose fit. A study by Richards (1981) indicated that physical changes with increasing age might be a cause of increasing fit problems of the shoulder length and bodice length in the older population. Chattaraman et al. (2013) found that an increase in age positively influenced the preference of looser fits by consumers and more body coverage by garments.

2.8.3 Psychosocial Factors (Body Image and Satisfaction)

It has also been reported that psychosocial factors such as body image and satisfaction or dissatisfaction (i.e., body cathexis) also have notable impact on consumers' fit preference and fit satisfaction (Alexander et al. 2005; Chattaraman and Rudd 2006; LaBat and DeLong 1990; Pisut and Connell 2007). A sign of this dissatisfaction is the hiding or camouflaging of body part/parts by wearing loose-fitting or baggy clothes (Chattaraman et al. 2013). Alexander et al. (2005) studied fit preference of young American female university students (aged 18 to 29) and found that women who were satisfied with their certain body parts, namely thigh, bust, hips, and waist, preferred a closer fit for those areas. Chattaraman and Rudd (2006) found that there is a correlation of lower body image and body cathexis with preference for greater body coverage through clothing by female consumers. In the case of male consumers, two psychological factors, namely drive for muscularity (DM) and social physique anxiety (SPA), also play a role in clothing fit preference (Chattaraman et al. 2013). Increased DM was found to be related to increased preference for more torso-revealing jeans.

2.9 FIT PREFERENCE SCALE

Attempts have been made to develop an apparel fit preference scale (AFPS) for use in research concerning consumers' fit preference. A non-comprehensive list of notable AFPS applied in fit preference research is presented in Table 2.1.

In order to measure the relationship between female consumers' preferences for apparel styling and their body image, Chattaraman and Rudd (2006) developed a 7-point semantic differential scale separately for tops and bottoms. Their scale included visual presentations of graded outlines of each styling attribute of each garment type on one graphic image of a female body. For example, they drew seven discriminatory outlines for side-seam silhouette sub-scaled between the bipolar adjectives of 'fitted' and 'unfitted', and numerically denoted as '1' for most fitted silhouette and '7' for fully unfitted silhouette. For ladies' tops, the scale measured participants' preferences in neckline depth (sub-scaled into bipolar adjectives of high and low), top length (long and short), sleeve length (full-length and sleeve-less), and

TABLE 2.1
Apparel Fit Preference Scale (AFPS) Developed for Research Purposes

Origin	AFPS	Used for	Products	Remarks
Chattaraman and Rudd (2006)	7-point semantic differential scale	Women	Tops and bottoms	2 generic scales for tops and bottoms
Alexander et al. (2005), Pisut and Connell (2007)	3-point graphical scale		Tops and bottoms	3 different drawings for each garment type
Chattaraman et al. (2013)	Multi-step industry relevant graphical scale	Men	Tops and bottoms	5 different drawings for each garment type

side seam silhouette (fitted and unfitted). Similarly for bottom items (trousers and skirt), their scale measured participants' preferences in looseness and tightness of overall fit (scaled into bipolar adjectives of fitted and unfitted), leg length (short and long), and waist level (high and low).

Instead of drawing graded outlines of styling attributes of a garment on one image, Alexander et al. (2005) and Pisut and Connell (2007) developed a 3-point scale that included three different drawings for each garment type representing the looseness and tightness of fit sub-scaled as fitted, semi-fitted, and loosely fitted. They presented similar scales for six garment types (dresses, jackets, blouses, skirts, pants, and jeans), which they employed to identify fit preference of female consumers in the USA.

Chattaraman et al. (2013) devised a multi-step industry-relevant graphical scale for measuring fit preference of male consumers based on the prevailing fit classification of garments. For example, they identified five different levels of fit for jeans, namely *skinny*, *slim*, *regular*, *relaxed*, and *loose*, through field study and thus developed a 5-point pictorial fit preference scale for jeans by visualising the garment outlines on body image by line diagram, using Abode illustrator software programme.

2.10 FIT EVALUATION

Fit testing is an integral part of clothing development and manufacture. The fit approval process is followed throughout the product development process and during and after production of garments (Kincade 2008). The fit evaluation method of a single garment belonging to a particular style at any stage throughout product development to production is related to, but someway different from, the fit evaluation methods followed in testing a complete garment sizing system. There are two different ways of fit testing to judge a garment sizing system. They are as follows:

1. Fit check of all garments representing all sizes of a garment sizing system on a selected group of fit models covering all body sizes of a particular anthropometric sizing table used to develop the garment sizes (Ashdown and O'Connell 2006; Bougourd 2007).
2. Fit check of one garment from each size on a number of participants representing the target population as a whole (Ashdown et al. 2004; Loker et al. 2005; Watkin 1995).

Fit testing is a must-do activity within the industry no matter if a single garment is tailor-made for a single customer or a style is mass-produced for a group of customers. However, there is no industry standard or protocol available for clothing fit evaluation, except for certain types of protective clothing, and as result, fit assessment methods vary among manufacturers and researchers (Ashdown and O'Connell 2006). In practice, clothing fit is evaluated by subjective or qualitative methods, although an objective or quantitative method would be more desirable.

2.10.1 Subjective Evaluation of Clothing Fit

Visual analysis is the main instrument of traditional fit evaluation. An expert or a panel of experts analyse fit by visually assessing a garment on a body. Either a live model or a dress form is used as a basis for testing the fit of garments. Although it is expensive to use live models (Yu 2004a), it returns substantial benefits due to the involvement of real human bodies that can express the wear comfort at static and dynamic positions. While testing with the live models, it is usually required for certain types of clothing, such as protective overalls, that the models perform certain body movements or follow a standard exercise protocol whilst wearing the clothing being tested (Huck et al. 1997).

The fit testing methods followed in the industry can be classified as the following two types:

1. Expert–Participant joint evaluation.
2. Expert-based evaluation.

The former one employs one or more live human models to try the garments to be tested, and fit experts jointly with the models evaluate the fit of garments. For a fit session, models are selected based on the representative size and shape of a company's target market. Some companies employ in-house fit models or select a member of staff who is of the right size to participate as a model in fit sessions when it is required. Some companies prefer to source them from specialist agencies (Jackson and Shaw 2001). A team of fit judges may include buyers, designers, pattern technicians, and clothing technologists. A fit model works together with the judges to check the fit issues by providing feedback on the comfort and ease during donning, doffing, sitting, bending, reaching, and walking (Bougourd 2007). The fit judges make decisions on the fit based on their visual analysis of the garment on the live model and the verbal report provided by the model. For general garment types, as already mentioned earlier, there is no standard protocol for running a fit session and no standardised exercise protocol for fit models to follow while working in a fit session. Only for a few types of protective clothing, there is a standardised exercise protocol available for fit models, such as ASTM F1154 for qualitative evaluation of fit of chemical protective clothing. As a result, the nature of fit sessions and their duration vary from company to company. Disadvantages associated with live models are their dimensional change in course of time and their unavailability in the required moment. It has been anxiously observed by Bye and LaBat (2005) that body measurements of fit models are not always checked prior to fit sessions.

The expert-based evaluation methods employ only dress mannequins or forms which are widely used by designers to create new designs through the draping technique. Previously available dress mannequins represented the size measurements of target customers only, not the shape very correctly. However, nowadays with the blessings of 3D body scanning and computer modelling technologies, more and more anthropometrically correct mannequins are available on the market, for example, FORMAX(R) models from CAD Ergomonics (Italy), TUKAform™ from Tukatech Inc. (USA), and AlvaForms from Alvalon Inc. (USA), etc. Like the former method with live models, there is also no standard protocol available to run a fit session with dress forms. This method solely depends on the judgement and expertise of the fit judges as it is not possible to get feedback from a dress form.

The subjective methods mentioned earlier are qualitative and include direct observation and description, participants' responses, and expert evaluations and ratings

(Watkin 1995). These methods are often not precise and well-communicated (Yu 2004a). However, the reliability of subjective evaluation of fit can be increased by using a panel of judges as opposed to a single judge, careful definition of the criteria for fit evaluation, and training of the judges prior to the assessment. It has been stated that capturing fit by still photographs or video camera instead of direct observation on the spot may also increase reliability of subjective fit evaluation methods (Kohn and Ashdown 1998; Petrova and Ashdown 2012). Applications of still photography, videography, and 3D body-scanning technology for the purpose of clothing fit evaluations are briefly discussed in the following paragraphs.

2.10.1.1 Photography for Fit Evaluation

Using the photography technique for clothing fit evaluation is seen in the work of Petrova and Ashdown (2012). For fit evaluation as a part of comparing different garment sizing systems, they took black-and-white photographs of front, right side, and back views of live models wearing test garments in their relaxed posture in standard conditions using a digital camera. They employed three expert judges to evaluate the fit of their test garments by traditional wrinkle analysis separately based on the photographs. However, this technique is not usually employed within the industry.

2.10.1.2 Video Technique for Fit Evaluation

Kohn and Ashdown (1998) applied video technique as a tool for clothing fit evaluation. They captured video image of live models wearing a test garment while turning slowly in a 360° circle. Later a panel of six judges viewed the video clips to evaluate the fit of the test garments. It was found that fit analysis using video image was as good as traditional fit analysis using a live model. The combination of video and Internet conferencing technologies already found industrial application in clothing fit analysis and has been described as beneficial in assessing fit of prototypes that are not on-site (Speer 2008).

2.10.1.3 3D Body Scanning for Fit Evaluation

3D body-scanning technology has been applied to capture an accurate representation of body/garment relation in several research works on clothing fit (Ashdown et al. 2004; Bye and McKinney 2010; Kohn and Ashdown 1998; Nam et al. 2005; Song and Ashdown 2010). Notable benefits have been seen in using the 3D body-scanning technique in fit analysis over still photographs and videotapes. 3D scan images can be rotated in any direction or enlarged to give a better view of any area, and using such images, a virtual fit analysis can be conducted at any time or location. Ashdown et al. (2004) used a VITUS/Smart 3D body scanner to capture fit images of a female model with pants on and processed those using Polyworks software for visualisation. They highlighted the benefits of employing 3D body-scanning technology for fit analysis; however, some limitations of this technique have also been reported.

Bye and McKinny (2010) used a combination of 3D body-scan and video techniques. Using a VITUS/Smart 3D body scanner from Human Solutions, they captured standing poses of live models and saved the scans as rotating movie files, which were later reviewed by judges for fit analysis. They reported that the fit judges had significant difficulties in evaluating certain fit criteria on 3D scans, and in some cases, judges' scores on 3D scans were not reliable. A similar study by Song and Ashdown (2010) captured 3D scans of clothed models and later processed those using Polyworks v7.2 software from Innovmetric Software Inc. before showing them to fit judges, who viewed the scans in Polywork/IMView 10.1 software to analyse the fit. This study also revealed that the fit analysis technique using 3D body scans suffered from lack of reliability in assessing fit and misfit at certain fit locations.

2.10.2 Guideline for Subjective Fit Evaluation

In several publications, five factors are mentioned as the standard criteria for the evaluation of fit, including grain, set, line, balance, and ease (Brown and Rice 2014; Erwin and Kinchen 1974; Kincade 2008; Stamper et al. 1988); each of them have a predefined meaning for this particular purpose. Erwin and Kinchen (1974) in their book *Clothing for Moderns*, of which the earliest print dates back to 1949, mentioned them as clues to a good fit. These are later widely adopted as the basic elements of clothing fit, and they provide a general guideline for subjective evaluation of the overall fit of a garment. These elements describe different aspects of fit; however, they are highly interrelated.

Lengthwise and crosswise grains represent the warp and weft of woven fabrics, respectively, or the wale and course of knit fabrics, respectively. In a well-fitted item of clothing, the lengthwise grains run parallel to the centre front or centre back, and crosswise grains run perpendicular to the centre front or centre back, unless they are cut on a diagonal or on the bias direction intentionally. 'Set' represents the smoothness and absence of any unwanted wrinkles. Wrinkles provide good clues to poor fit. Wrinkles which are not a part of a design, for example, the folds from gathers, tucks, and flare introduced by designer, are termed as 'set wrinkles' (Brown and Rice 2014) or 'fitting wrinkles' (Liechty et al. 2010). These cannot be eliminated by ironing; they simply reappear when fabrics are relaxed. The type of these wrinkles and their location hint at the cause of fit problems in garments. There are two general types (tight and loose) and three forms (horizontal, vertical, and diagonal) of these wrinkles, and the presence of any of these can indicate different fit problems, including insufficient, excess, or misplaced ease (Brown and Rice 2014; Liechty et al. 2010).

'Line' indicates the silhouette, seam lines, or any design lines, which in well-fitted clothing should be in agreement with the body contour and curves. 'Balance' stands for the symmetrical nature of a garment when split into equal halves. Unless intentional asymmetry is introduced, symmetry represents well-fitted clothing. Ease in clothing refers to the free space between the body and the garment when it is being worn. Presence of adequate ease in clothing determines the level of fit.

Based on these factors or elements, several guidelines or rules for good fit of clothing have been described in Brown and Rice (2014), Liechty et al. (2010), McDevitt (2009), and Rasband and Liechty (1994). They are commonly followed by professionals and researchers in industry and academia for defining and evaluating clothing fit.

2.10.2.1 Fit Evaluation Scale (FES)

Several fit evaluation scales (FES) have been developed for use in research works (see Table 2.2). However, no such scale has been so far developed for industrial

TABLE 2.2
Fit Evaluation Scales (FES)

Origin	FES	Ranking/Points System	Remarks
Petrova and Ashdown (2012)	4-point scale	Rank 1 = best fit Rank 4 = worst fit	Overall fit in bust-waist-hip area was checked by 3 judges.
Bye and McKinny (2010)	5-point scale	Score 1 = unacceptable fit Score 2 = poor fit Score 3 = acceptable fit Score 4 = good fit Score 5 = excellent fit	Pant fit based on 17 criteria and dress fit based on 24 criteria were checked by 6 judges
Song and Ashdown (2010)	5-point scale	Score 3 = good fit	Jacket fit at 18 critical body locations was judged by 3 judges
Ashdown and O'Connell (2006)	4-point scale	No scoring system, rather descriptive options	Jacket fit at 13 specific areas and an overall fit assessment was done by 9 judges
Nam et al. (2005)	5-point scale	Score 5 = strongly agree Score 1 = strongly disagree	Fit of cooling vest at 36 torso areas checked by 3 judges
Ashdown et al. (2004)	3-point scale	Score 1 = acceptable Score 0 = marginal Score –1 = unacceptable	3 judges evaluated pant fit using 15 criteria
Shen and Huck (1993)	9-point scale	Score 0 = best fit Other scores range from –4 to +4	25 fit criteria of female bodice were checked by 3 judges

application. The notable fit scales published in the recent literature are described in the following paragraphs in their reverse chronological order of development.

4-Point Scale by Petrova and Ashdown (2012)

A 4-point scale, with rank 1 being the best and 4 being the worst, was described in Petrova and Ashdown (2012). Their research focused on the comparison of four different garment sizing systems, and as a part of it, three judges evaluated the fit of test jackets from different sizing systems by viewing the black-and-white photographs of live models with jackets on. No specific criteria for fit evaluation were employed; rather, the judges ranked the overall fit of each jacket prototype in the bust-waist-hip area which were relevant for the sizing systems being compared.

5-Point Scale by Bye and McKinny (2010)

Bye and McKinny (2010) developed a 5-point scale for use by a panel of six judges to evaluate pant fit based on 17 criteria and dress fit based on 24 criteria by viewing the videotapes of body-scan data. The weight of scores in this scale was presented as: 1 for unacceptable fit, 2 for poor fit, 3 for acceptable fit, 4 for good fit, and 5 for excellent fit. The evaluation criteria for pant and dress were further divided into three broad groups based on the following considerations:

1. Overall alignment (O/A): how perfectly a garment is aligned on wearer's body at centre front, centre back, left-side seam, right-side seam, waist, shoulder seam, sleeve and hem of the dresses, and at inseams, left-side seam, right-side seam, front crotch seam, and back crotch seam of the pants.
2. Dart placement (DP): direction and placement of front bust darts, front skirt waist darts, back shoulder darts, back bodice waist darts, and sleeve darts of the dresses, and of front darts and back darts of the pants.
3. Looseness/tightness (L/T): amount of ease at front neckline, back neckline, bust, bodice back, front waist, back waist, front hip, back hip, armscye, and sleeve of dresses, and at waist band, stomach, front crotch, front full hip, front thigh, back crotch, back upper hip, back full hip, and back thigh of pants.

5-Point Scale by Song and Ashdown (2010)

Song and Ashdown (2010) designed a 5-point scale for judges to evaluate fit in live session and also in 3D scans of clothed models. In their scale, the middle point of the scale, 3, represented good fit. Using this scale, the three judges assessed the fit and misfit of a collarless V-neck princess style jacket at 18 critical body fit locations, namely bust, waist, hip, cross chest, cross back, sleeve bicep, underarm, neck side point, neck back point, neck front point, shoulder seam placement, side seam placement, hem balance, centre front seam placement, centre back seam placement, princess line placement, sleeve length, and hem length. The judges utilised five basic elements of fit (ease, line, grain, balance, and set) described by Erwin et al. (1974) and Leichty et al. (2010) to define the good fit.

4-Point Scale by Ashdown and O'Connel (2006)

Ashdown and O'Connell (2006) devised a 4-point scale for fit evaluation of a semi-fitted, unlined, collarless, hip-length jacket. This scale did not include any numerical scoring system, rather it asked the judge to use any of the four descriptive statements covering good fit to poor fit for each criterion. Using this scale, nine judges evaluated jacket fit on 14 criteria focussing on 13 specific garment areas and an overall fit assessment.

5-Point Scale by Nam et al. (2005)

In order to analyse the fit of liquid cooled vest prototypes, Nam et al. (2005) employed three judges who were asked to indicate the extent of their agreement or disagreement with the goodness of fit at 36 locations (16 front locations, 12 back locations, and 8 side locations) using a 5-point response scale. The responses were scored as '5 for strongly agree' and '1 for strongly disagree'.

3-Point Scale by Ashdown et al. (2004)

Ashdown et al. (2004) formulated a 3-point scale for evaluation of pants fit using 3D body-scanning technology. A panel of three judges involved in their study individually analysed the pant fit based on 15 criteria that included 13 fit locations in pants and overall fit at front and back. Scoring system in this scale featured as 'acceptable (1), marginal (0), or unacceptable (–1)' with consideration of ease, line, balance, set, and grain. Fit locations that were considered for pant fit in addition to overall front and overall back fit were waist front, waist back, waist placement front, waist placement back, abdomen front, abdomen back, crotch front, crotch back, below buttocks, thigh front, and thigh back.

9-Point Scale by Shen and Huck (1993)

Shen and Huck (1993) described a subjective scale that allowed the judges to mark a score from −4 to +4 on 25 fit criteria of a female bodice, where a zero indicated the best possible fit for each criterion. They grouped the fit criteria in three categories as follows:

1. Overall fit covering bust line circumference, waistline circumference, shoulder seam position, side seam position, strain/excess ease at shoulder tip, shoulder seam length, side seam length.
2. Bodice front fit covering strain/looseness at bust level, bodice front length, strain/looseness at upper chest (above bust level), front neckline position, ease at front neckline, gapping/strain at front armhole, armhole shape at bodice front, shoulder dart position at front, waist dart position at front.
3. Bodice back fit covering bodice back length, strain/looseness at shoulder blade level, strain/looseness above shoulder-blade level, neckline position at back, ease at back neckline, gapping/strain at back armhole, armhole shape at back, shoulder dart position at back, waist dart position at back.

For each of these criteria, nine different markings were possible, ranging from one extreme of the fit criteria to the opposite extreme; for example, 'much too tight = score −4' to 'much too loose = score +4' in case of a circumference; or 'much too short = score −4' to 'much too long = score +4' in case of length measurement; or 'far too high or far towards the front = −4' to 'far too high or far towards the back = +4' in case of position of any particular feature. Chen (2007) also utilised this scale to evaluate the fit of basic garments for Taiwanese young women.

2.10.2.2 Wearer Rating Scale (WRS)

Wearer rating scales (WRS) can be used to understand the perceived fit by the live models. Usually, a WRS is found to be applied for the fit evaluation of protective clothing, not for any regular type of clothing. Nam et al. (2005) utilised a 5-point response scale for the fit analysis of liquid cooled vest prototypes. They asked their models to answer nine questions after wearing the vest prototypes they designed. The first four questions were designed to get information on wearing comfort, flexibility, tightness, and movement restriction perceived by the subjects. The next two questions were asked about ease during donning and doffing the prototypes and about fit adjustment. Questions 8 and 9 addressed the wearer's preference on the vest style and design. For each question, the possible responses were scored as 1 = strongly disagree, 2 = disagree, 3 = neutral, 4 = agree, and 5 = strongly agree.

Huck et al. (1997) developed a 9-point wearer acceptability scale for the fit evaluation of protective overalls by employing live models. Their models followed an exercise protocol as per ASTM F1154–88 with the overalls on before completing the feedback form that included nine numerical responses ranging from 1 to 9 on 13 fit and comfort criteria. The scale featured two bipolar adjectives for each criterion with possible marking from 1 to 9, where 1 means 'the worst' and 9 means 'the best' in terms of fit and comfort. The first three criteria dealt with how the subjects felt after wearing the test garments, and the rest of the criteria addressed fit and design issues of the prototypes.

2.10.3 Objective Evaluation of Clothing Fit

Attempts have been made to develop objective methods for the evaluation of clothing fit. Yu (2004b) summarised five approaches to the objective evaluation of clothing fit, including moiré optics, algebraic mannequin, waveform, pressure mechanics, and computer modelling of fit. It is difficult to achieve objective evaluation of clothing fit, and a limited number of research works have been carried out in moiré optics, algebraic mannequin, waveform, and pressure mechanics for objective fit evaluation (Yu 2004b). However, the area of virtual simulation of clothing fit has been well exploited in the last three decades, and a number of 3D clothing CAD systems that support virtual clothing prototyping and fit simulation are already available on the market (Sayem et al. 2010). It is further discussed separately in the next section.

2.11 VIRTUAL CLOTHING FIT ANALYSIS

The subjective fit evaluation practices described in Section 2.10.1 are only applied at the post-manufacture stage of a garment prototype or a batch of garments during bulk production. The virtual fit analysis technique provides an opportunity to review and forecast the clothing fit at the pre-manufacture stage of clothing. This offers both subjective and objective analysis of clothing fit by combining a visual check of the simulated fit within the computer screen together with the numerical analysis of stress between fabric and body, and ease mapping from virtual clothing (Sayem et al. 2010; Lim and Istook 2001). Notable 3D CAD systems that support virtual clothing prototyping and fit simulation and are available on the market are Vstitcher (Browzwear), Accumark 3D (Gerber/Lectra), Clo 3D (Clo), Modaris 3D (Lectra), TUKA3D (Tukatech), 3D Runway (OptiTex), Style3D (*Linctex*), and Vidya (Assyst). As a 3D virtual prototyping solution, they associate 2D patterns, material information, and 3D virtual mannequins. They enable simulation of 3D design from 2D pattern pieces developed by a wide range of 2D CAD software with the help of physically based simulation engines. They also come with built-in libraries of avatars, fabrics, trims, and accessories. Users can pick any available avatar from the library to morph it to the required size and shape by means of a range of parameters, from age and gender, through body measurements and posture, to skin tone and hair style, and even through the stages of pregnancy to represent the target customers. Once digital pattern pieces are ready, they can be placed around an avatar and virtually stitched as a preparation of virtual simulation. To reproduce the realistic drape of fabrics, it is essential to assign relevant mechanical properties of fabrics before initiating virtual simulation. While built-in libraries of fabrics and other related materials together with their mechanical characteristics are available within these systems, it is also

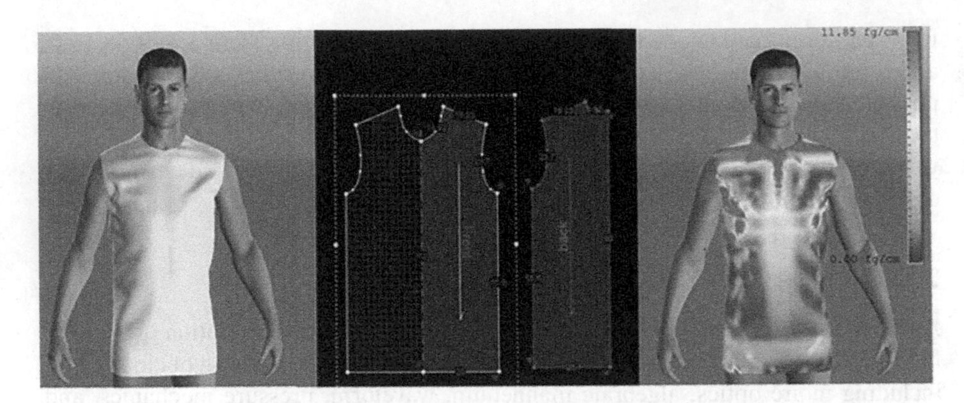

FIGURE 2.2 Virtual fit simulation (left) from 2D pattern pieces (middle) and tension mapping (right) in a commercial CAD system.

possible to input new fabric properties taken from an objective fabric measurement system, such as KES (Kawabata Evaluation System), FAST (Fabric Assurance by Simple Technique), and others related to software systems and their suppliers. Once the simulation is done, the CAD systems allow their users to review fit of simulated garments on accurately sized virtual avatars and facilitate virtual designs to be communicated with any remote partner via the Internet platform. This reduces the requirement of frequent physical prototypes to be made and travelled from one place to another, and thus can shorten the product-development lead-time and the associated costs (Ernst 2009; Sayem et al. 2010). Figure 2.2 represents a virtual fit simulation of a short sleeve men's shirt and its corresponding tension map generated for the purpose of virtual fit analysis.

Although the virtual analysis technique shows many potentials for improving the present scenario of prevailing fit problems within the industry, it cannot be seen as a complete alternative to the physical fit evaluation processes. The technology is still waiting to find a widespread application across the industry, covering both retailers and manufacturers, even though the commercial systems have been available on the market for a notable period of time.

Within the environment of a 3D CAD system, it is possible to carry out a 360-degree review of virtual prototypes to check for wrinkles and bulges. In addition, it is also possible to run different maps like tension, stretch, and ease maps to get an idea of the tightness and looseness of the garments on virtual avatars. However, it is reported that such visual review is still a subjective approach and does not provide enough clues to make error-free decisions if any mass production can be initiated without seeing a physical prototype (Kim 2009; Lim 2009; Kim and LaBat 2013). A true objective approach to virtual fit analysis was presented by Sayem (2017). He utilised the numerical value of relevant virtual drape parameters—virtual tension (gf/cm), stretch (%), and pressure (dyne/cm^2 or gm/cm^2) —to analyse the virtual drape of ladies' blouses and recommended a concept of an intelligent 'virtual fit prediction system' which is based on the principle of colour matching system currently in use in the textile industry.

2.12 CONCLUSION

Clothing fit is a function of three constituent components, namely anthropometry, manufacture, and comfort. Clues to address the problems of misfit also lie in these areas. Consumers' preferences in clothing fit vary depending on several demographic, physical, and psychological factors. Although the concept of clothing fit has been well-established in several textbooks, and several fit evaluation scales have been used in several research works, a universal scale and a standard protocol for industrial application is still missing. Virtual clothing prototyping and fit analysis is a recent development in the field of clothing technology and is catching a notable degree of attention. It provides an opportunity of combining subjective and objective fit evaluation at the design stage of clothing, but it is not expected to be a complete alternative to the physical fit analysis process followed at the post-manufacture stage. Rather it may significantly reduce the number of physical prototypes and needs of fit sessions, if the available 3D CAD technology can be widely implemented throughout the industry, covering both designers and manufacturers.

REFERENCES

Abend, J. 1993. "Our fits over fits." *Bobbin* 34 (1): 78–79.

Alexander, M., L.J. Connell, and A.B. Presley. 2005. "Clothing fit preferences of young female adult consumers." *International Journal of Clothing Science and Technology* 17 (1): 52–64.

Ashdown, S.P., S. Loker, K. Schoenfelder, and L. Lyman-Clarke. 2004. "Using 3D scans for fit analysis." *Journal of Textile and Apparel Technology and Management* 4 (1): 1–12.

Ashdown, S.P., and E.K. O'Connell. 2006. "Comparison of test protocols for judging the fit of mature women's apparel." *Clothing & Textile Research Journal* 24 (2): 137–146.

Bougourd, J. 2007. "Sizing systems, fit models and target markets." In *Sizing in Clothing Developing Effective Sizing Systems for Ready-to-Wear Clothing*, edited by S.P. Ashdown, 108–151. Cambridge, England: Woodhead Publishing Ltd.

Brown, P., and J. Rice. 2014. *Ready-to-Wear Apparel Analysis*, 4th ed. Columbus: Pearson.

Bubonia, J.E. 2014. *Apparel Quality: A Guide to Guide Evaluating Sewn Products*, 97–101. England: Fairchild Books/Bloomsburry Publishing Inc.

Bye, E., and K. LaBat. 2005. "Analysis of apparel industry fit sessions." *Journal of Textile and Apparel, Technology and Management* 4 (3): 1–5.

Bye, E., and E. McKinney. 2010. "Fit analysis using live and 3D scan models." *International Journal of Clothing Science and Technology* 22 (2/3): 88–100.

Cain, G. 1950. *The American Way of Designing*. New York: Fairchild Publications.

Chattaraman, V., and N.A. Rudd. 2006. "Preference for aesthetic attributes in clothing as a function of body image, body cathexis and body size." *Clothing Research Journal* 24: 46–61.

Chattaraman, V., K.P. Simmons, and P.V. Ulrich. 2013. "Age, body size, body image and fit preference of male consumers." *Clothing Research Journal* 31 (4): 291–305.

Chen, C.-M. 2007. "Fit evaluation within the made-to-measure process." *International Journal of Clothing Science and Technology* 19 (2): 131–144.

Chi, L., and R. Kennon. 2006. "Body scanning of dynamic posture." *International Journal of Clothing Science and Technology* 18 (3): 166–178.

Cilley, J. 2016. Apparel & Footwear Retail Survey Report, Solving the Fit Problem. New York: Body Labs.

Das, A., and R. Alagirusamy. 2010. "Garment fit and comfort." In *Science in Clothing Comfort.* New Delhi: Woodhead Publishing Limited.

Das, A., and R. Alagirusamy. 2011. "Improving tactile comfort in fabric and clothing." In *Improving Comfort in Clothing*, edited by G. Song, 216–244. Cambridge: Woodhead Publishing Limited.

DesMarteau, K. 2000. "Let the fit revolution begin." *Just-style.com*, October 16. Accessed October 2, 2022. www.just-style.com/analysis/let-the-fit-revolution-begin_id92196.aspx.

Devarajan, P., and C. Istook. 2004. "Validation of female figure identification technique (FFIT) for apparel software." *Journal of Textile and Apparel, Technology and Management* 4 (1): 1–11.

Ernst, M. 2009. "CAD/CAM powerful." *Textile Network* 4: 20–21.

Erwin, M.D., and L.A. KinChen. 1974. *Clothing for Moderns*, 5th ed. London: The Macmillan Company.

Giovis, J. 2007. "More fitting clothes urged." *Chicago Tribune*, January 22. Accessed October 2, 2022. http://articles.chicagotribune.com/2007-01-22/business/0701220053_1_fit-technologies-apparel-industry-long-torso.

Gupta, D. 2014. "Anthropometry and the design and production of apparel: An overview." In *Anthropometry, Apparel Sizing and Design*, edited by D. Gupta and N. Zakaria. Cambridge: Woodhead Publishing Limited.

Ho, C.P., J. Fan, E. Newton, and R. Au. 2011. "Improving thermal comfort in apparel." In *Improving Comfort in Clothing*, edited by G. Song, 165–181. Cambridge: Woodhead Publishing Limited.

Huck, J., O. Maganga, and Y. Kim. 1997. "Protective overalls: Evaluation of clothing design and Fit." *International Journal of Fashion Design, Technology and Education* 9 (1): 45–61.

Istook, C.L. 2008. "Three-dimensional body scanning to improve fit." In *Advances in Apparel Production*, edited by C. Fairhurst, 94–116. England: Woodhead publishing limited.

Jackson, T., and D. Shaw. 2001. *Mastering Fashion Buying and Merchandising Management.* Hampshire and London: Palgrave Macmillan.

Johnson, K.K.P. 1990. "Impressions of personality based on body forms: An application of Hillestad's model of appearance." *Clothing & Textile Research Journal* 8 (4): 34–39.

Kawabata, S., and M. Niwa. 1989. "Fabric performance in clothing and clothing manufacture." *Journal of Textile Institute* 80 (1): 19–50.

Kawabata, S., M. Niwa, and Y. Yamashita. 2002. "Recent development in the evaluation technology of fiber and textiles: Toward the engineered design of textile performance." *Journal of Applied Polymer Science* 83: 687–702.

Kincade, D.H. 2008. *Sewn Product Quality, a Management Perspective*, 229–259. New Jersey, USA: Pearson Education.

Kim, D. 2009. *Apparel Fit Based on Viewing of 3D Virtual Models and Live Models* (Unpublished doctoral Thesis). The University of Minnesota. Retrieved from the University of Minnesota Digital Conservancy, https://hdl.handle.net/11299/54510.

Kim, D.-E., and K. LaBat. 2013. "An exploratory study of users' evaluations of the accuracy and fidelity of a three-dimensional garment simulation." *Textile Research Journal January* 83 (2): 171–184.

Kirchdörfer, E., and M. Rupp. 2007. "3D-body-scanning-SOA technology-available data and analysis for additional applications." *D1.3—Towards Mass-Customisation, 4th Technical Edition (Last Revision)*, NMP2-CT-2004-511017, LEAPFROG CA, 5–27. Accessed June 17, 2015. www.leapfrog-eu.org/leapfrog-ca/upload/resultDocuments/L-CA_D1.3%20TE4-Virtual%20%20v2.0.pdf.

Kirk, W., and S.M. Ibrahim. 1966. "Fundamental relationship of fabric extensibility to anthropometric requirements and garment performance." *Textile Research Journal* 36 (1): 37–47.

Koblyakova, E.B. 1980. *Osnovikonstruirovaniyaodejdi (Basics of Clothing Construction)*. Moskva, USSR: Legkayaindustriya. (cited in Petrova and Ashdown, 2008).

Kohn, I.L., and S.P. Ashdown. 1998. "Using video capture and image analysis to quantify apparel fit." *Textile Research Journal* 68 (1): 17–26.

LaBat, K.L., and M.R. Delong. 1990. "Body cathexis and satisfaction with fit of apparel." *Clothing and Textile Research Journal* 8 (2): 43–48.

Li, Y. 2001. "The science of clothing comfort." *Textile Progress* 21 (1–2): 1–35.

Liechty, E., J. Rasband, and D.P. Steineckert. 2010. *Fitting & Pattern Alteration, a Multi-Method Approach to the Art of Style Selection, Fitting, and Alteration*, 2nd ed. New York: Fairchild Books.

Lim, H.S. 2009. *Three Dimensional Virtual Try-on Technologies in the Achievement and Testing of Fit for Mass Customization* (Unpublished doctoral Thesis). North Carolina State University. Retrieved from http://www.lib.ncsu.edu/resolver/1840.16/3322.

Lim, H.S., and C.L. Istook. 2001. "Drape simulation of three-dimensional virtual garment enabling fabric properties." *Fibers and Polymers* 12 (8): 1077–1082.

Loker, S., S. Ashdown, and K. Schoenfelder. 2005. "Size-specific analysis of body scan data to improve apparel fit." *Journal of Textile and Apparel, Technology and Management* 4 (3): 1–13.

Manuel, M.B., L.J. Connell, and A.B. Presley. 2010. "Body shape and fit preference in body cathexis and clothing benefits sought for professional African-American women." *International Journal of Fashion Design, Technology and Education* 3 (1): 25–32.

McCullough, E.A., B.W. Jones, and P.J. Zbikowski. 1983. "The effect of garment design on the thermal insulation values of clothing." *ASHRAE Transactions* 89 (2A): 327–352.

McDevitt, P.J.M. 2009. *Complete Guide to Size Specification Technical Design*. New York: Fairchild Books.

Mckinnon, L., and C.L. Istook. 2002. "The effect of subject respiration and foot positioning on the data integrity of scanned measurements." *Journal of Fashion Marketing and Management* 6 (2): 103–121.

Moll, J.M.H., and V. Wright. 1972. "An objective clinical study of chest expansion." *Annals of the Rheumatic Diseases* 31: 1–8.

Nam, J., D.H. Branson, H. Cao, B. Jin, S. Peksoz, C. Farr, and S. Ashdown. 2005. "Fit analysis of liquid cooled vest prototypes using 3D body scanning technology." *Journal of Textile and Apparel, Technology and Management* 4 (3): 1–13.

Niwa, M., M. Nakanishi, M. Ayada, and S. Kawabata. 1998. "Optimum silhouette design for ladies' garments based on the mechanical properties of a fabric." *Textile Research Journal* 69 (8): 578–588.

Petrova, A., and S.P. Ashdown. 2008. "Three-dimensional body scan data analysis: Body size and shape dependence of ease values for pants' fit." *Clothing and Textiles Research Journal* 26 (3): 227–252.

Petrova, A., and S.P. Ashdown. 2012. "Comparison of garment sizing systems." *Clothing and Textiles Research Journal* 30 (4): 267–284.

Pisut, G., and L.J. Connell. 2007. "Fit preference of female consumers in the USA." *International Journal of Clothing Science and Technology* 11 (3): 366–379.

Postle, R. 1983. "Objective evaluation of the mechanical properties and performance of fabrics and clothing." In *Objective Evaluation of Apparel Fabrics, Proceeding of 2nd Australian-Japan Symposium, Melbourn, 1983*, edited by R. Postle, S. Kawabata, and M. Niwa. Osaka: Textile Machine Society of Japan.

Rasband, J., and E. Liechty. 1994. *Fabulous Fit Speed Fitting and Alteration*, 1st ed. New York: Fairchild Publications.

Richards, M.L. 1981. "The clothing preferences and problems of elderly female consumers." *The Gerontologist* 21 (3): 263–267.

Sayem, A.S.M. 2017. "A novel approach to fit analysis of virtual fashion clothing." IFFTI Annual Conference 2017, The Amsterdam Fashion Institute (AMFI), Amsterdam.

Sayem, A.S.M., R. Kennon, and N. Clarke. 2010. "3D CAD systems for the clothing industry." *International Journal of Fashion Design, Technology and Education* 3 (2): 45–53.

Sheldon, W.H. 1954. *Atlas of Men, a Guide for Somatotyping the Adult Male at All Ages.* New York: Harper.

Shen, L., and J. Huck. 1993. "Bodice pattern development using somatographic and physical data." *International Journal of Clothing Science and Technology* 5 (1): 6–16.

Simmons, K.P., C. Istook, and P. Devarajan. 2004. "Female figure identification technique (FFIT) for apparel, Part II: Development of shape sorting software." *Journal of Textile and Apparel, Technology and Management* 4 (1): 1–14.

Simmons, P., and C. Istook. 2003. "Comparing 3D body-scanning and anthropometric methods for apparel applications." *Journal of Fashion Marketing and Management* 7 (3): 306–332.

Song, H.K., and S.P. Ashdown. 2010. "An exploratory Study of the validity of visual fit assessment from three-dimensional scans." *Clothing and Textiles Research Journal* 28 (4): 263–278.

Speer, J.K. 2008. "Victoria's secret: Framing the fit problem." *Apparel*, November 2. Accessed July 24, 2015. http://apparel.edgl.com/news/Victoria-s-Secret-Framing-the-Fit-Problem 64260.

Stamper, A.A., S.H. Sharp, and L.B. Donnell. 1988. *Evaluating Apparel Quality*, 214–229. New York: Fairchild Publication.

Watkin, S. 1995. *Clothing the Portable Environment*, 2nd ed. Iowa: Iowa State University Press.

Yoo, S. 2003. "Design elements and consumer characteristics relating to design preferences of working females." *Clothing and Textiles Research Journal* 21 (2): 49–62.

Yu, W. 2004a. "Subjective assessment of clothing fit." In *Clothing Appearance and Fit: Science and Technology*, edited by J. Fan, W. Yu, and C. Hunter, 30–42. England: Woodhead Publishing Limited.

Yu, W. 2004b. "Objective evaluation of clothing fit." In *Clothing Appearance and Fit: Science and Technology*, edited by J. Fan, W. Yu, and C. Hunter, 72–88. England: Woodhead Publishing Limited.

Ziegert, B., and G. Keil. 1988. "Stretch fabric interaction with action wearable: Defining a body contouring pattern system." *Clothing Research Journal* 6 (4): 54–64.

3 The Virtual Fitting Process—How Precisely Does 3D Simulation Represent Physical Reality?

Simone Morlock, Christian Pirch and Anke Klepser

CONTENTS

3.1 INTRODUCTION

The fashion industry is in a state of upheaval. It is moving away from the traditional product development process towards virtual prototyping. This requires the design

DOI: 10.1201/9781003264958-5

and product development processes to be made more efficient and also more sustainable. Technologies such as 3D clothing simulation enable a clear reduction in product development times from the design idea to the point of sale (Istook and Hwang 2001; Daanen and Hong 2008; Ernst 2009; Krzywinski and Siegmund 2017; Sayem 2019; Morlock 2020a).

Nevertheless, switching from the traditional development processes to virtual prototyping is no small undertaking. Newcomers in particular are faced with a number of different challenges, which are not apparent at the start and therefore frequently underestimated. In addition to high investment costs and the necessary familiarisation time for these complex systems, there are other hurdles to overcome.

Before introducing 3D technologies to a company, various questions must be answered; for example, where in the supply chain should 3D be used? In the design, in the pattern, during trying on, in the sales and/or marketing? Who is the user? The designer, the pattern maker, or the 3D fashion artist? What are the individual prerequisites, and what infrastructure is already available? What is the goal? Visualisation or virtual try-on and fit analysis?

This last point is particularly important in this regard. As a matter of principle, it is necessary to differentiate between product visualisation and virtual try-on because, not only are the goals different for these two, but the principles for successful application of the 3D simulation also differ.

3D visualisation is used for product communication in design, sales, and marketing purposes. It supports the design process and facilitates rapid design decisions and modifications. A purposeful decision can be made for or against a design very early on in the collection development process effectively without the prototypes. The 3D product model makes communication along the entire value chain more transparent and eliminates misunderstandings, including in communication with suppliers, that may result from traditional 2D sketches (Morlock 2020c). In contrast to 3D visualisation, 2D sketches are often not clear and lead to different interpretations. In addition, online shops and other marketing and sales media can be provided with product visualisations without having to manufacture a single garment. This not only makes clothing development more efficient, but also reduces the associated carbon footprint of physical production and transport.

In contrast, in the virtual try-on, the appearance of the clothing is also taken into account, but always in relation to the fit. It is used to check the silhouette and proportions, as well as the fit, and if necessary, to make adjustments. Therefore, it is possible to approve an item without seeing a physical prototype. The greatest challenge here is that the 3D user must have technical understanding of the simulation algorithms of the systems as well as expertise in traditional fit and pattern (Morlock 2020b). Without this expertise, a reliable virtual try-on cannot be carried out. Furthermore, there are numerous other factors to take into account to enable a reliable simulation, which represents more than just the visual appearance. For example, the virtual workmanship processing does not always correspond to the one in real-life, and vice versa: the real-life processes cannot be transferred 1:1 to the virtual world (Zangue et al. 2020).

The research presented here represents the bases and guidelines for a high-quality and reliable fit simulation and assessment. The 3D systems Vidya

Assyst (Assyst GmbH), VStitcher from Browzwear (Browzwear Solutions Pte Ltd), and CLO 3D (CLO Virtual Fashion LLC) are used. However, the findings presented here are system-independent. This means that the findings can also be applied to systems from Optitex, Lectra, Gerber, TUKA3D, and others. Furthermore, the potential and the limits of virtual technology for fit assessment are reviewed, and differences between real-life and virtual product development are discussed.

3.2 STATUS QUO FOR VIRTUAL TRY-ON

The use of 3D simulation software for clothing has been considered from a scientific perspective since the 1980s, although a true clothing simulation has only been possible since the 1990s (Volino et al. 2005). Initially, the available 3D CAD software were mainly used by enthusiasts and visionaries. Yet, there were still major teething problems to be resolved. The technology was clearly limited; for example, the material could not be adequately simulated. The users lacked specialist knowledge and, ultimately, the acquisition costs were very high (Power 2013). However, in recent decades, simulation systems have undergone significant development and various programs such as Vidya, VStitcher, Clo3D, Optitex, and a few more are available on the market. As a result, 3D clothing simulation has been the subject of intensive scientific investigation and widely discussed within the industry (Jevšnik et al. 2017; Lapkovska and Dabolina 2018; Lee et al. 2007; Song and Ashdown 2015). Most research works initially focussed on individual products. Thereby, clothing parts were developed using 3D and then finished in 2D, or 2D patterns were simulated in 3D. The simulation of textiles was a major challenge. Consequently, many research projects focussed on fabric drape. Only a few projects looked at the avatars used in the systems. In this case too, the focus was on the comparison between the real body and the virtual avatar, and the restrictions placed by the 3D simulation programs (Sayem 2019; Klepser and Pirch 2021; Balach et al. 2020). Some research works also looked at the generation of avatars outside of the systems (Pirch et al. 2021; Klepser and Pirch 2021; Rudolf et al. 2021; Leipner and Krzywinski 2013; Kozar et al. 2014; Zhang and Krzywinski 2019).

Numerous studies looked at the accuracy and efficiency of the 3D simulation programs and compared virtual products with real products. These were, for example, skirts (Lapkovska and Dabolina 2018; Lee et al. 2011; Wu 2011), shirts (Sayem 2017), pants (Kim and LaBat 2013; Lee et al. 2007; Song and Ashdown 2015), diving suits (Naglic et al. 2016), etc. (Ögülmüs et al. 2015; Porterfield and Lamar 2017; Apeagyei and Otieno 2007; Lee and Park 2017). Both consistency and difference were reported. The focus was on fabric draping and the corresponding material characteristics, such as bending, stretching, weight, smoothness, etc. (Lim and Istook 2011; Lee et al. 2011; Sayem 2017). Fit conformities and differences were investigated using fit locations, body shapes, and overall fit. Required extra widths were analysed with 3D simulation programs (Lage and Ancutiene 2017). The use of system-internal analysis maps was also taken into account from a scientific perspective (Liu et al. 2017; Brubacher et al. 2021).

3.3 THE THREE PILLARS OF SIMULATION

Irrespective of the specific 3D systems mentioned earlier, three bases are always needed for creating a garment simulation. These are the pattern, the material, and the avatar (Sayem 2019; Morlock 2020b, 2020c, 2020a). For the virtual try-on, it is essential that a verified well-fitting pattern is used and physical and mechanical parameters of textile material are digitised and used in the system. In addition, the avatar must correspond to the customer in dimensions and shape. If one of these parameters is disregarded, the achieved virtual fit will not correspond to real-life. Thus, the fit assessment based on the virtual products will be unreliable and non-reproducible.

3.3.1 PATTERN

The traditional simulation of clothing starts with the process of sewing digital pattern pieces virtually. Therefore, 3D simulation of clothing is not possible without a pattern. The more proven and well-fitting the pattern, the more reliable the resulting simulation. This particularly presents a significant challenge to companies, who no longer create their own patterns and only work with suppliers' patterns. For this reason, basic and style pattern libraries are recommended. These ensure the fit, avoid recurrent and avoidable fit problems, and enable the creation of quick and efficient simulations.

It is important to note that the patterns must be prepared for the 3D simulation, and thus they differ from production-ready patterns. No seam allowance is required in pattern pieces for 3D simulation. Darts and pocket positions should be drawn in with internal cutting lines and not above drill holes. In production daily routine, there are clothing pieces that are not part of the pattern and not graded, for example, neckline cuffs. This is different for the virtual sewing process; here they need to be graded and included in the pattern.

3.3.2 MATERIAL

The properties of textile material impact the fit and visual appearance of clothing products. This applies to the 3D simulation of clothing, too. They determine the virtual drape behaviour of the textile in the simulation. For this reason, the testing of material parameters and input of data in the systems is of crucial importance to the simulation result. The current challenge for individual 3D users is that until now, there have not been any international standards for the digitisation of material parameters, and each software has its own requirements covering what and how to test. Even the required units differ. For example, the bending parameter is entered as "μNm" for one system, "Bend W/L (***dyn*/cm²**)", for another, and "Bending weft/warp ((g*mm²)/(s^2*rad))" for a third system. Furthermore, the bending for Vidya, for example, must be tested according to the cantilever principle, whereas the test specification of CLO 3D is the measurement of the so-called contact distance and length.

There are different ways to digitise the fabric parameters (Power 2013; Sayem 2017; Lim and Istook 2011). One possibility is the use of the test devices from the software providers. But these devices are linked to one specific software and not universally applicable. For this reason, a universal approach for digitising material parameters is required. Therefore, a universal digitisation process has been developed at Hohenstein, which is based on the requirements of the 3D systems and on current standards. The five most important parameters, which impact the virtual drape behaviour and therefore must be digitised for prevailing CAD systems, were identified: weight, thickness, elongation, bending, and fold volume (Coalition 2020; DTB—Dialog. Trust. Business. The fashion and textile community 2019; Kuijpers et al. 2020; Kuijpers et al. 2021; Morlock 2021).

The weight is measured according to DIN EN 12127:1997 (CEN—Europäisches Komitee für Normung 1997), the thickness according to DIN EN ISO 5084:1996 (CEN—Europäisches Komitee für Normung 1996), the bending stiffness according to DIN 53362: 2003–10 (DIN—Deutsches Institut für Normung e.V. 2003), and the stretch according to DIN EN ISO 13934–1:2013–08 (DIN—Deutsches Institut für Normung e.V. 2013). The Drapeometer test is conducted according to DIN 54306:1979–02 (DIN—Deutsches Institut für Normung e.V. 1979). Minor adaptions of the standards are required for some parameters; for example, the input of fabric bending property in CLO 3D, amongst others. All measured values can then be calculated and converted for each 3D software system so that there is a universal application of the virtual material parameters in all systems.

A reproducible digitisation process must be considered as important so that reliable material parameters can be determined for the virtual try-on. However, the 3D user must be aware that, due to the different calculation algorithms and mesh structures of 3D systems, there may be simulation differences when comparing systems. A reproducible digitisation process significantly helps to minimise such differences.

In addition to the physical parameters of textile material, the textures are required to describe a textile surface fully. The textures represent the surface finish of a textile. Yet, they only visualise the optical properties and thus are not required for the virtual try-on. Moreover, for virtual try-on it is recommended to simulate without texture and colour. Both may influence the fit evaluation to that effect that a comparable assessment is made more difficult or even lead to wrong fit evolution results.

3.3.3 Avatar

Like in reality, correct fitting models are an important basis for fit testing in the virtual world. Avatars represent the customers and target group of each company. The body measurements and body shape, as well as the body posture, influence the fit (Zangue et al. 2022). For this reason, it should be viewed as essential that an item of clothing is simulated and assessed on an avatar, which correctly represents the realistic human body shape in every clothing size.

For the visualisation, it is important that the appearance of the avatar is as idealised and visually appropriate as possible. In contrast, for the try-on, a technical avatar is needed whereby the focus is on real body proportions, for example, realistic and non-idealised stomach shapes.

There are different kind of avatars that can be used for virtual try-on. These are system-internal avatars, virtual fit dummies, scanatars, and customised avatars.

Each system provides its own standard avatars, which can be adjusted to specific size requirements by setting parameters. The parameter settings differ between systems and are sometimes limited. In particular, the adjustment of body shapes to suit defined customer groups is not fully feasible. Furthermore, although the system avatars are visually appropriate, they are sometimes very idealised and do not always represent the real body shapes of people, especially not in larger sizes. The consequence of this is that the fit of simulated clothing can, in some cases, end up being different than it would be with real clothing on a living model.

In addition, there is an option to work with fit dummies. Some of the known dummy suppliers offer physical and virtual dummies. The advantage of these is that the identical dummy bodies enable a reproducible fit comparison in both real and virtual formats. However, compared to the human body, a dummy body is always idealised, which in turn influences the fit of the clothing worn by real people.

The alternative is to generate a "scanatar". People, for example, in-house models, can be scanned in the 3D body scanner. The 3D dataset needs to be postprocessed with regard to the software requirements and imported to the 3D systems.

Generally, customised avatars can also be created. Hohenstein has developed fitting avatars for fitting processes, which are based on representative body data from size surveys (Hohenstein Institute and HUMAN SOLUTIONS GmbH 2007; Morlock 2015; Morlock et al. 2009, 2020) from the last 20 years. These specifically represent realistic body shapes in all sizes and different target groups. They are equipped with so called fit lines. Examples are shown in Figure 3.1. These avatars also have a rig (skeleton) to create different postures.

Fit lines are necessary because digital garments can be put on any position on the body by the software user, and avatars cannot give feedback in regard to a comfortable garment position like a real person. Without these aids, a reproducible positioning of the garment is not possible.

Simulated movement and generation of different poses is the subject of much scientific research and discussion (Rudolf et al. 2021; Feng et al. 2015; Jolly et al. 2019; Harrison et al. 2018; Krzywinski and Siegmund 2017). Yet, there are still a number of technical hurdles in research to be resolved so that the fit during movement can be reliably assessed virtually on an avatar. This starts with the size accuracy and the realistic deformation of the body and soft tissue of the avatars (Brake et al. 2022). Certainly, the use of 4D scanning systems and the knowledge of the body surface in motion will lead to optimised avatars in different postures and movement (Klepser and Pirch 2021; Klepser et al. 2021; Klepser et al. 2020; Klepser 2021).

3.4 FUNDAMENTALS OF FIT TESTING

The fit assessment always starts with the style description and fit definition. The fit definition is an essential component of the fit assessment. It defines widths, lengths, and the position of the clothing; for example, the rise on trousers. The waistband shape, collar, and detail widths and depths must also be described. This must be considered for both the real-life and virtual fitting processes. In addition, it has to be defined in which way the garment is worn and which function is required. Finally, an exact description of the processing is necessary.

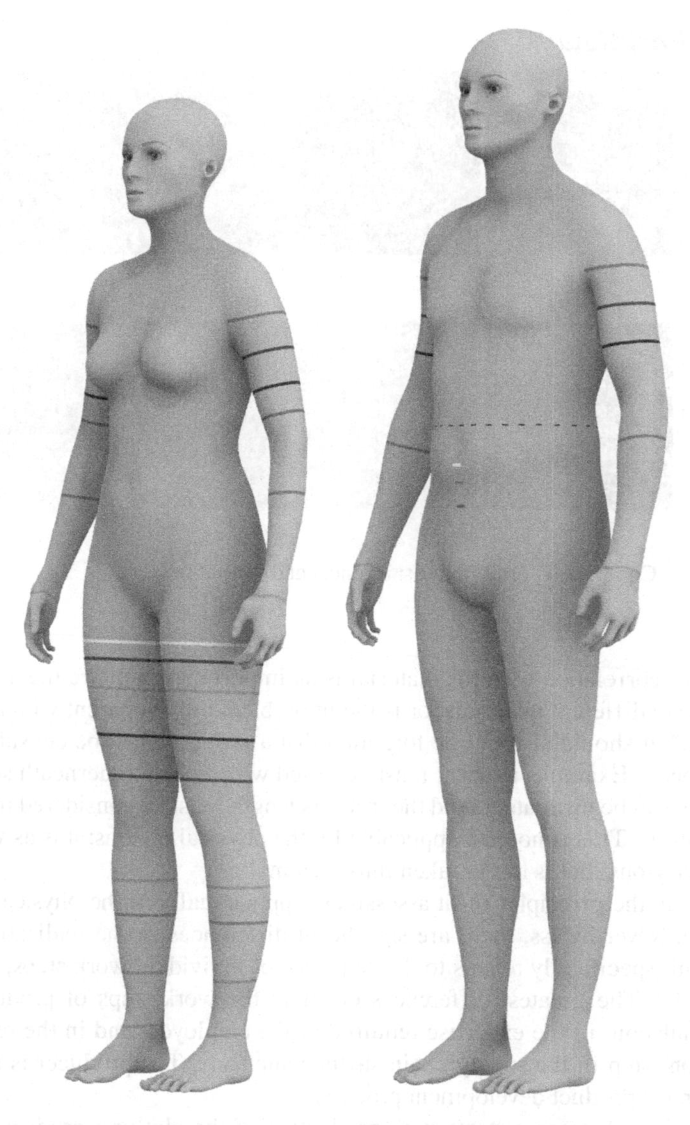

FIGURE 3.1 Hohenstein avatars with fit lines, female German size 38 (bust girth 88,0 cm), male German size 50 (chest girth 100 cm).

But what is a fit problem? To be able to objectively assess a fit, the framework conditions of a fit test must be defined in a fit protocol. In general, there is a fit problem if there are constrictions, an unwanted accumulation of material, diagonal folds, tension, and/or irregular seam contours. If a problem occurs, it must be checked in each case whether this can only be identified on this individual person or on other fit models with the same clothing size and figure type. It should as well be investigated whether this problem occurs only in one particular size or in all clothing sizes. It is helpful to check whether the fit problem can be identified on the fit dummy, too. In

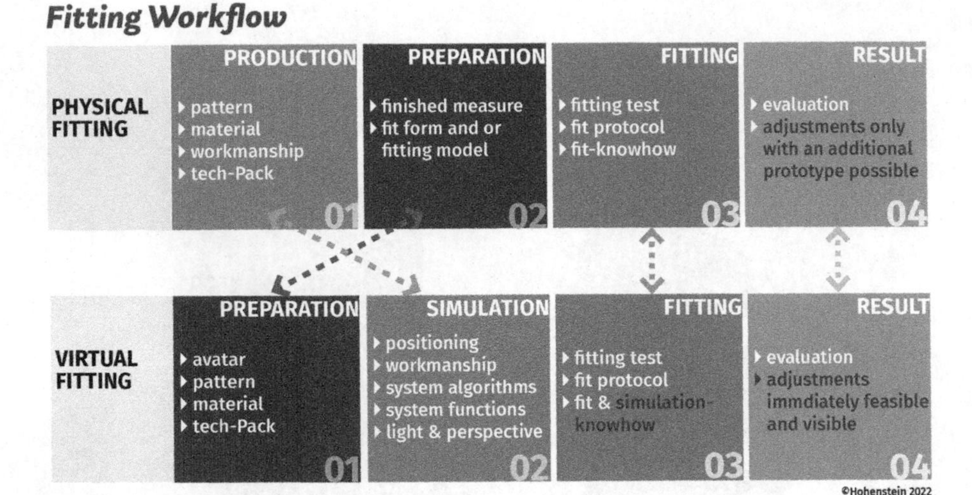

FIGURE 3.2 Comparison of fit process: physical and virtual try-on.

addition, the correlation with the material is an important point: are the differences attributable to different materials, or is the fit problem only apparent with a particular material? It should also not be forgotten that a product must be considered as a complete outfit. Example: a jacket must be tested with a shirt underneath so that the width ratios can be estimated. And the trousers length must be considered in connection with shoes. This is not just applicable for the physical try-ons; it is as well valid for virtual try-ons, but is rarely taken into account.

In general, the principles of fit assessment apply equally to the physical and virtual try-on. Nevertheless, there are significant differences in the realisation of the try-ons. This specifically relates to the sequence of individual work steps, as shown in Figure 3.2. The greatest differences occur in the work steps of production, or rather, simulation, in the expertise required of the employee, and in the result. The "Production" step in the supply chain shifts completely. The producer is no longer involved in the product development process.

The traditional process starts with production of the clothing product, then the try-on is prepared by checking the finished measurements and organising suitable dummies and/or the fit model. If problems are identified during the fit assessment, these cannot be resolved immediately, and, if there is any doubt, another prototype must be produced. In contrast, the virtual process starts with the preparation of the simulation, which includes, for example, digitisation of the material parameters. The simulation itself corresponds to production in the broadest sense. Yet, far more expertise is required for the simulation than for production. The employee carrying out the task must have technical knowledge of sewing clothing, but also be proficient in the specific system functions and algorithms of virtual sewing. When the simulation is completed, the try-on can take place without further preparation. It is necessary to have 3D knowledge in addition to fit expertise for the virtual fit assessment because the system algorithms can influence the virtual fit. This significantly increases demands on the employee concerned. Unlike the real-life fitting, necessary changes and optimisations resulting from the fit assessment can be implemented immediately and as often as required.

In summary, it can be concluded that a basic requirement for conducting virtual try-ons is the modification of traditional work processes in product development. The traditional product development process from design idea and pattern creation to production of the prototype and try-on cannot be transferred to the 3D process directly. If this is not taken into account, it is not possible to carry out reliable try-ons.

3.5 FACTORS INFLUENCING THE FIT OF CLOTHING PRODUCTS

The factors influencing the fit of a clothing product are varied and cannot always be assessed objectively, particularly if taking into account individual preferences of the consumer how to wear a garment. Nevertheless, for the fit process, it is important to note that the virtual fit of clothing is influenced by more factors than the physical fit.

3.5.1 Factors Influencing the Physical Fit

In general, the following factors influence the fit of clothing: target group (which defines both the requirements of the clothing as well as the body shape and measurements), clothing sizes, pattern geometries, material properties, workmanship, required function, and required comfort, as well as the defined quality level. However, the predominant fashion also plays an important role, of course. All factors together result in an optimum product.

3.5.2 Factors Influencing the Virtual Fit

In addition to the factors that affect the real-life fit, the virtual fit is influenced by system functions, system algorithms, and individual user behaviour. 3D simulations can be changed and consciously manipulated. The difference between "idealised" and "realistic" is demonstrated in Figure 3.3. It shows a simulation of a slim fit jacket

FIGURE 3.3 Idealised (A) and realistic (B) 3D simulation of a slim fit jacket in comparison.

with the same pattern and material parameters but in two versions. In contrast to the idealised simulation on the left (A), the realistic simulation (B) clearly shows that significant fit problems are apparent in the displayed jacket.

The differences are down to the avatars and the application of other system functions. With the idealised version, a dummy without arms is used, whereas the realistic simulation uses an avatar with realistic body shape. The idealised version only simulates the jacket without shirt and trousers. This does not indicate that there is no additional width in the waist and hip area. Another crucial difference is apparent in that the idealised simulation actively used system functions for the idealisation. These include, for example, a pin and iron functions which enable the user to eliminate each crease. A reliable fit assessment is no longer possible from this point in time. In addition to discussed aspects, the virtual fit is influenced by many other technical factors which are described in the following sections. Therefore, for the reproducible implementation of virtual try-on, it must be considered important to define a fit protocol, which describes and defines the use of system functions and parameters in addition to the usual fit parameters. This will ensure a reliable and comparable virtual fit result. Based on this finding, Hohenstein has developed the so-called user guidelines for the reproducible virtual fit test.

3.6 USER GUIDELINES—THE BASIS FOR THE VIRTUAL FIT TEST

The user guidelines standardise and simplify the processes for creating a virtual prototype and therefore regulate digital cooperation within the company and in the value chain. Simulation results can be considered objectively, and unconscious changes and manipulations can be recognised and avoided. The user guidelines ensure that all parties involved in product development, including the suppliers, use the same bases for creating the simulation, and keep to the defined simulation protocol. They represent an important quality assurance tool in the virtual value chain. The following aspects are included in the guidelines:

- Pattern, material, avatars.
- Garment positioning on the avatar, for example, with the help of fit lines.
- Virtual workmanship variants and processing sequence.
- Camera lighting setup—setting and perspective.
- System functions: application, sequence, idealisation tools.
- Resolution of mesh structure.
- Analysis of fit assessment maps.

3.6.1 Pattern, Material, Avatars

The parameters of pattern, material, and avatars are the bases for each simulation and thus form the three basic pillars. As described in Section 3.1, for the virtual try-on it is necessary to ensure that reliable and proven patterns are used to digitise the material parameters, and that the avatar is generated in accordance with the specific customer group. It must also be ensured that all persons involved in product development use the same bases because all three pillars influence the fit. As in reality,

FIGURE 3.4 Material influence on product appearance, hoodie simulated in canvas and jersey.

material properties have an impact on the virtual fit. Figure 3.4 shows the influence of the physical parameters of textile material on the appearance of a hoodie as an example. Comparing the two, the hoodie in jersey and in canvas show clear differences. Each demonstrates a different drape behaviour and folds. This becomes particularly obvious on the drape of the hood.

Figure 3.5 presents the significant influence of body shape on fit. Both avatars wear a t-shirt based on the same pattern and material parameters. The avatars have the same body measurements but differ in body shape and posture. The difference

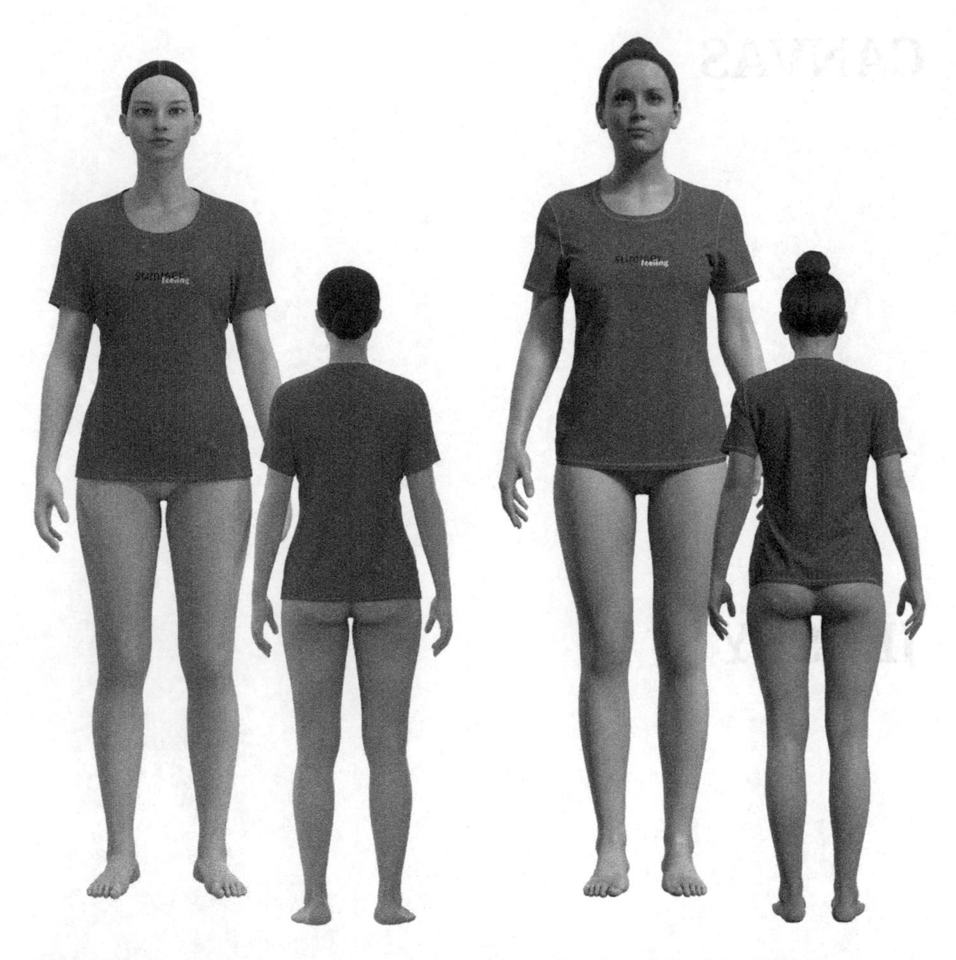

FIGURE 3.5 Influence of avatar geometry on the fit of a t-shirt, left CLO 3D avatar in German size 38, right Vstitcher avatar in German size 38 (Bust girth 88,0 cm).

in posture alone leads to a different fit. Assessing the two fit results, the left version would be rated as good. The fit of the shirt on the right side would be rated worse because of the folds in the bust area and back. The poor fit is not caused by the pattern.

3.6.2 Positioning on the Avatar

The positioning of the garment on the avatar has a major influence on the simulation and the result. In contrast to reality, in the digital world avatars do not give feedback to the wearing comfort and correct location of the garment. In fact the garment can be positioned at any location on the body. Especially challenging is the location of the waistband. Here the belly bottom is the only orientation point to locate the pants on low waist, mid waist, or high waist. Each 3D CAD user will put on the garment differently. Therefore, even though the pattern, the avatar, and the material parameters are the same, the result of the virtual try-on will differ because the relocation of the garment has an impact on the fit. Therefore, reference lines have to be defined to ensure reproducible garment positions on the avatar. Therefore, Hohenstein developed so called fit lines (see Section 3.3.3). They enable each 3D programme user and supplier in the value chain to generate a comparable garment simulation.

3.6.3 Virtual Workmanship Variants and Processing Sequence

In contrast to real-life production, where the garment is finished in several steps using different machines, in the virtual production, the garment is sewn directly on the body. The sequence of the single processing steps can and sometimes needs to be different, compared to the processing steps in reality. The advantage of the virtual sewing process is that it can be repeated without the material being damaged, hence the result can be optimised through several iterations. Though, the major disadvantage is that the processing of seams, the seam allowances as well as the seam type, have no influence on the virtual fit result. As example: in reality a fell seam influences the crotch seam of trousers because multiple textile layers are joined which causes tension. But tension cannot be identified in the simulation. Similarly, the seam allowances in the hem are not taken into account in the simulation, although in reality they must be cut in to avoid tensions. Therefore, the influence of the processing and workmanship within the virtual fit test cannot be checked. Yet, the virtual processing can have individual settings to correspond to the physical processing. For example, the hem can appear double-layered while it's single-layered just by visually adjusting the thickness. The order of the processing steps is individual depending on the product, and depending on the amount of adjustment used in the system for the seams, the simulation result can appear more realistic or less. Nevertheless, the demand for virtual processing should work as close to reality as possible. The best method to achieve this aim is to define workmanship guidelines. Otherwise, the result of the virtual fit can be falsified.

3.6.4 Camera Lighting Setup—Setting and Perspective

The evaluation of the fit based on the simulation can be done in two ways. On the one hand, the fit can be checked based on the 3D object itself or based on images of it. The advantage of checking the 3D object is that the user can change the perspective continuously, whereas the fit check based on the images is bound to the present perspective. In addition, the light setting influences the visual appearance of the images and can thus influence the evaluation of the fit. For a consistently good evaluation of fit based on images, standard settings for light and the perspective need to be defined.

The light setting in the simulation affects the colour appearance as shown in Figure 3.6, as well as the visibility of the shape and folds of the garment. Depending on the setup, folds appear clearly visible and even dramatic or are almost invisible. The blouse to the left shows significant folds, therefore a fitting problem would be assessed. In addition, the pattern would be optimised. In contrast, the blouse on the

FIGURE 3.6 Influence of light on the product appearance.

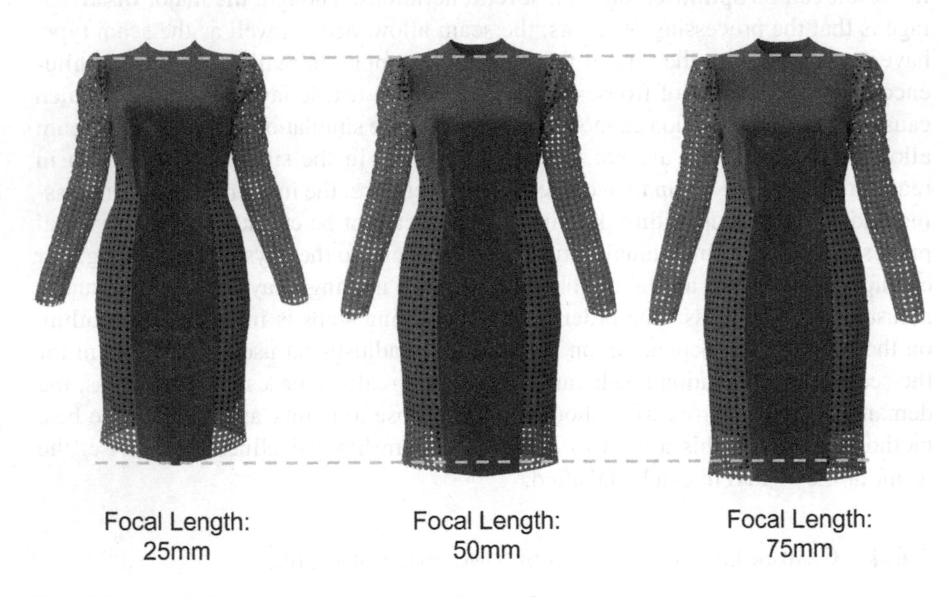

FIGURE 3.7 Influence of camera perspective on fit assessment.

right picture side seams to have simply little folds. Actually, it is the same blouse captured with differing lighting factors.

For the assessment of the fit, it is important to avoid textures and colours when creating the images, as these distract from the folds and may also make them unrecognisable. This may lead to incorrect fit evaluations. Furthermore, a lighting setup should be chosen that evenly illuminates the garment and no areas are over-lit or, on the contrary, too dark.

Another issue is to define the camera perspective in order to avoid incorrect fit evaluation results. Figure 3.7 shows different renderings of the same dress, where three different focal lengths are used. A line is positioned on the shoulder and hem which enables a comparison of the perspective distortion of the dress. The distortion due to camera perspective is most apparent with a focal length of 25 mm. The hem is significantly lower in the front area than in the back. With focal length of 50 mm, the balance still seams not correct. The rendering with focal length of 75 mm provides a visualisation on which the balance can be assessed as realistic.

The assessment of the fit in the system itself on the 3D object is not dependent on the factor's perspective and light, as these can be adjusted at any time. Yet, images to validate the fit are very dependent on these factors, and incorrectly chosen settings can lead to errors. Therefore, the settings used to create the images must be chosen correctly, applied consistently, and reproducible across the products.

3.6.5 System Functions: Application, Sequence, Idealisation Tool

The user of 3D systems has the option to influence the simulation result in more ways than simply sewing and positioning. 3D systems include tools, such as those for ironing, placing needles, and setting internal lines and their fold angles. The purpose of these tools is to recreate the workmanship processes used in real-life as authentically as possible. An exact sequence for virtual processing and the extent to which tools are used beyond sewing is different for the individual garment and user. The consequence of this may be that the tools lead to an over-idealisation of the simulation result, which is sufficient for the visual requirements, but which distorts the fitting result since wrinkles or other supposedly unwanted effects are dissolved virtually by pins and ironing, but cannot be dissolved in real production. Therefore, the degree to which the garment is processed by these features also has a great influence on the result of the simulation and thus strongly affects the virtual fit test. Therefore, for carrying out the virtual fit test, it is recommended not to use these idealisation tools in order to ensure that fit problems can be seen and analysed.

To assess garment fit, the 3D programme user has to accept wrinkles and folds and must not smoothen the simulation. The evaluation of the garment fit is only possible if the realistic fit is simulated and not manipulated. The use of system functions can be defined by the user guideline, too.

3.6.6 Resolution of Mesh Structure

The geometries used in the simulation are represented by polygon meshes, whereby the surface models are discretised by vertices. The mesh density has a significant impact on the simulation speed as well as on a virtual fit result. The greater the number of points to be processed, the longer the processing time. Even though, in the approximation of the pattern geometry curves described by functions, a higher mesh density means a more realistic representation, a greater point density for the simulation geometry means that the drape behaviour and physical textile properties can be represented more realistically. The mesh density of the avatar has a crucial impact on the visual display, as well. If the mesh density of the simulation geometry is too high in relation to the mesh density of the avatar, it causes the avatar surfaces to be pushed through the clothing in the regions where the clothing lies on the body. The left avatar in Figure 3.8 shows a low-resolution mesh which leads to a non-realistic representation of the trousers' shape, because the faces are depicted on the trousers. With a much higher resolution mesh, as seen in Figure 3.8 on the avatar on the right, these errors disappear.

Therefore, mesh resolution has an impact on the visual impression of the simulation as well as changes the fit. Like the calculation of the curves for the pattern geometry, it is linked to the accuracy of the measurements. The lower the resolution, the more incorrect the body shape and dimensions will be. The lower the resolution, the less folds will occur (Pirch et al. 2021).

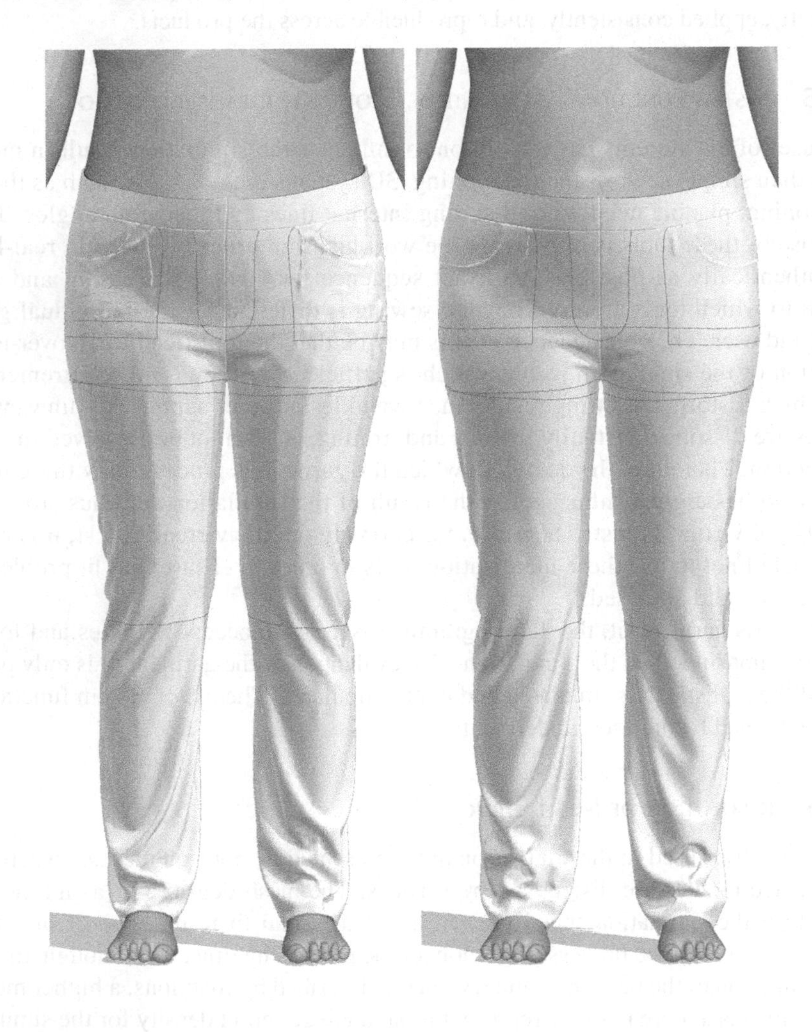

FIGURE 3.8 Impact of mesh resolution on trousers appearance.

3.6.7 Analysis Maps for Fit Assessment

There is also an option to use the analysis maps provided by the systems to assess the virtual fit of clothing. Within these maps, the calculated data from the simulation is displayed in heat maps on the simulation geometries or avatars. The calculation of realistic results using the analysis maps are based on, in addition to realistic avatars, the physical parameters of textile material. These data can be used to calculate three factors: the tension in the textile, the distance to the body, and the pressure applied to the body by the clothing. The results can then be tested with regard to the requirements of the garment in an analytical virtual try-on. A particular advantage is that these analysis maps go beyond the subjective assessment and provide numerical results for the virtual try-on. The calculated values are displayed in a colour gradient on the clothing. In most cases, the regions with values at the upper end of the range are shown on the garment in red and very low values in green. However, red values do not necessarily mean that the virtual fit is poor. Within which size range the calculated values are assigned is up to the user. The user has to choose the values according to the material. The values must be adjusted to the values identified from the testing process. In relation to the result of the virtual fit test, the interpretation of the heat maps has to be learned. Just as in the work with the idealisation tools, the results must be put into relation and their cause must be explored, whether the origin is from the material or the pattern, as described in Figure 3.3.

3.7 THE THREE STAGES OF VIRTUAL TRY-ON

To determine the basis for a standardised and qualitative virtual fit test, it is necessary to define the basic requirements and data described in the earlier section. Furthermore, all relevant factors for assessing the fit must be already carried out in the simulation. This includes, for example, the corresponding shoes on the avatar when practising the fit of a pair of trousers. In addition, the garment must be simulated in the corresponding layer, that is, the other layers must be prepared. For fit evaluation, the complete outfit needs to be simulated. The actual testing of the simulation results is then split into two phases: the first is the visual try-on and the second is the analytical try-on. There is no representation of texture in either of these tests as creases and shape can be better identified and are more visible, as demonstrated in Figure 3.9 on the avatar in the middle. In the visual fit test, the focus is on the shape of the simulations; here, the contours, width, and length of the garments are tested against the design and fit specifications that have been defined. The surface is examined for creases and transverse pulls. During the analytical try-on, the structural connections of the simulation grid are then examined using the analysis maps of the system as shown in Figure 3.9 on the avatar on the right. For the final design check, texture and colours can be added, as shown on the left avatar in Figure 3.9. But it can be seen that the folds and wrinkles disappear after adding the texture, so the fit cannot be evaluated anymore.

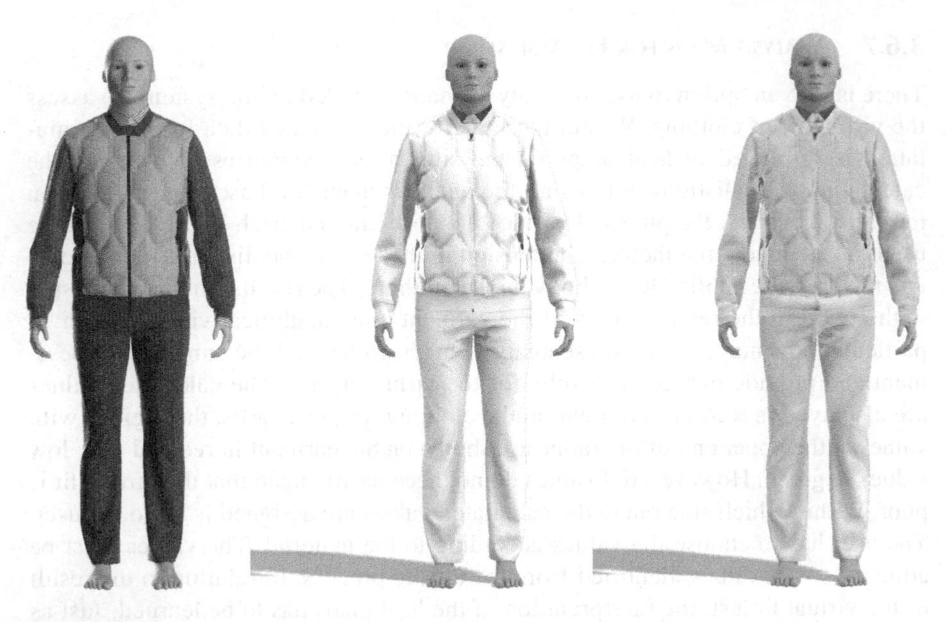

FIGURE 3.9 From left to right, the final simulation, visual and analytical inspection.

3.8 CONSTRAINTS OF VIRTUAL TRY-ON

3D Simulation systems already offer many possibilities to realistically simulate garments based on real material parameters and to carry out a fit test, though there are still limitations in the simulation of the avatar, the garment, and the verification of the fit. The current state system avatars do not provide soft tissue. Therefore, the body garment interaction cannot be demonstrated realistically, although there are other 3D programme solutions to investigate garment pressure on soft body areas (VitalMechanics). Furthermore, body shape in movement, especially the representation of the joints, needs optimisation. In the simulation software, the plastic deformation is not taken into account because the deformation of the material is usually considered linear. In this way, a material can be stretched seemingly infinitely. Only by referring to the values from the material test can an irreversible distortion be determined. Moreover, these values are not stored within the simulation, and the material will always return to its original state. The same applies to the processing: seams are indestructible and do not tear under high stress in the simulation. Therefore, the simulation of the seam does not influence the virtual fit in contrast to reality. Buttons and other rigid ingredients on the clothing show no stress limit, as well. In reality, material characteristics and workmanship have a high influence on the fit as well as on wearing comfort, which can only be tested to a limited extent in the simulation. For example, a fell seam in direct body contact can be noticeably uncomfortable and can cause unwanted wrinkles and folds in reality, but in the simulation, it is only visually represented and therefore has no influence on the simulation and the evaluation of the virtual fit. And furthermore, the mistakes in tomorrow's production cannot be simulated in today's simulation either.

The final wearing comfort of the garment depends not only on fit but also on skin-sensorial perception, which cannot be evaluated in the garment simulation. Therefore, the biggest limitation of the simulation systems is that the garment can only be tested for fit purposes, such as style, silhouette, proportion, and significant fit issues. The feeling of touching and wearing material can only be generated in reality. Despite these limitations, 3D simulation software supports the virtual product development. Yet, to ensure a reliable evaluation of the virtual garment fit, the constraints are to be considered by the user.

3.9 CONCLUSION

There are many different ways to process a digital garment, and there are in parts significant differences to reality. The factors influencing the virtual fit test are manifold, and, as shown, clothing simulation is still very complex. Yet, the visualisation of garments can be already performed on high level. Virtual fit testing is feasible for selected products and helps to reduce development times and prototypes.

For successful implementation of the 3D simulation programme in the current processes, a combination of technological understanding for the simulation algorithms as well as traditional fit and pattern know-how is essential. A goal-oriented virtual result can be ensured and the benefits of the technology fully utilised only by connecting these topics. Furthermore, it is recommended to define user guidelines in order to ensure reliable simulations in the 3D value chain. This means to rule the basics as well as the system-specific specifications, like pattern, avatars and material, positioning on the avatar, digital workmanship, lightning, camera and rendering settings, mesh resolution, and the use of idealisation tools of the software. Following well-defined user guidelines, 3D simulation programs represent garments close to reality and lead to smarter, more cost effective, and more sustainable product development.

However, unresolved systemic and technical issues relating to all stages of the simulation need to be considered by the user. The decisive factors are the implementation of consistent avatars representing realistic body shapes of target groups, the lack of soft tissue avatars, and the lack of standards for the digitisation of material parameters. Other major issues are the limitation of workmanship simulation, since there is no influence on the fit, and the limitation of the drape simulation based on the material parameters, since it is not always close to reality.

The knowledge concerning the strengths and weaknesses of the 3D garment simulation systems helps to interpret the digital fit result correctly and to realise a successful virtual prototyping.

REFERENCES

Apeagyei, Phoebe R., and Rose Otieno. 2007. "Usability of pattern customising technology in the achievement and testing of fit for mass customisation." *Journal of Fashion Marketing and Management: An International Journal* 11 (3): 349–365. doi: 10.1108/13612020710763100.

Balach, Monika, Agnieszka Cichocka, Iwona Frydrych, and Marc Kinsella. 2020. "Initial investigation into real 3D body scanning versus avatars for the virtual fitting of garments." *Autex Research Journal* 20 (2): 128. https://doi.org/10.2478/aut-2019-0037.

Brake, Elena, Yordan Kyosev, and Katerina Rose. 2022. "3D garment fit on solid and soft digital avatars-preliminary results." *Communications in Development and Assembling of Textile Products* 3 (2): 97–103. doi: 10.25367/cdatp.2022.3.p97-103.

Brubacher, Kristina, David Tyler, Phoebe Apeagyei, Prabhuraj Venkatraman, and Andrew Mark Brownridge. 2021. "Evaluation of the accuracy and practicability of predicting compression garment pressure using virtual fit technology." *Clothing and Textiles Research Journal*: 0887302X21999314. doi: 10.1177/0887302X21999314.

CEN—Europäisches Komitee für Normung. 1996. EN ISO 5084:1996 Bestimmung der Dicke von Textilien und textilen Erzeugnissen. In EN ISO 5084:1996.

CEN—Europäisches Komitee für Normung. 1997. DIN EN 12127:1997 Bestimmung der flächenbezogenen Masse. In DIN EN 12127:1997.

Daanen, Hein, and Sung-Ae Hong. 2008. "Made-to-measure pattern development based on 3D whole body scans." *International Journal of Clothing Science and Technology* 20 (1): 15–25. doi: 10.1108/09556220810843502.

DIN—Deutsches Institut für Normung e.V. 1979. DIN 54306:1979-02 Prüfung von Textilien; Bestimmung des Fallvermögens von textilen Flächengebilden (Norm zurückgezogen).

DIN—Deutsches Institut für Normung e.V. 2003. DIN 53362:2003-10 Prüfung von Kunststoff-Folien und von textilen Flächengebilden (außer Vliesstoffe), mit oder ohne Deckschicht aus Kunststoff—Bestimmung der Biegesteifigkeit—Verfahren nach Cantilever.

DIN—Deutsches Institut für Normung e.V. 2013. DIN EN ISO 13934-1:2013-08 Textilien—Zugeigenschaften von textilen Flächengebilden—Teil 1: Bestimmung der Höchstzugkraft und Höchstzugkraft-Dehnung mit dem Streifen-Zugversuch (ISO 13934-1:2013).

DTB—Dialog. Trust. Business. The fashion and textile community. 2019. "Arbeitskreis Digitalisierung physikalischer Materialparameter." Accessed January 4, 2022. www.dialog-dtb.de/events/arbeitskreis-digitale-materialparameter/.

Ernst, Michael. 2009. "CAD/CAM powerful." *Textile Network* 4: 20–21.

Feng, Andrew, Dan Casas, and Ari Shapiro. 2015. "Avatar reshaping and automatic rigging using a deformable model." Proceedings of the 8th ACM SIGGRAPH Conference on Motion in Games, Paris (FRA).

Harrison, Darcy, Ye Fan, Egor Larionov, and Dinesh K. Pai. 2018. "Fitting close-to-body garments with 3D soft body avatars." 9th International Conference and Exhibition on 3D Body Scanning and Processing Technologies, Lugano (CH).

Hohenstein Institute and HUMAN SOLUTIONS GmbH. 2007. "SizeGERMANY." Hohenstein Institute and HUMAN SOLUTIONS GmbH Accessed January 3, 2022. www.sizegermany.de/.

Istook, Cynthia L., and Su-Jeong Hwang. 2001. "3D body scanning systems with application to the apparel industry." *Journal of Fashion Marketing and Management: An International Journal* 5 (2): 120–132. Doi: 10.1108/EUM0000000007283.

Jevšnik, Simona, Zoran Stjepanovič, and Andreja Rudolf. 2017. "3D virtual prototyping of garments: Approaches, developments and challenges." *Journal of Fiber Bioengineering & Informatics* 10: 51–63.

Jolly, Kanika, Sybille Krzywinski, P.V.M. Rao, and Deepti Gupta. 2019. "Kinematic modeling of a motorcycle rider for design of functional clothing." *International Journal of Clothing Science and Technology* 31 (6): 856–873. Doi: 10.1108/IJCST-02-2019-0020.

Kim, Dong-Eun, and Karen LaBat. 2013. "Consumer experience in using 3D virtual garment simulation technology." *The Journal of the Textile Institute* 104 (8): 819–829. doi: 10.1080/00405000.2012.758353.

Klepser, Anke. 2021. "Using 3D 4D technology to investigate workwear fit in motion (oral presentation)." 9th European Conference on Protective Clothing (ECPC), Stuttgart (GER).

Klepser, Anke, Angela Mahr-Erhardt, and Simone Morlock. 2021. Grundlagenuntersuchung zur Erschließung der 4D-BodyScanner-Technologie für die Analyse bekleidungsbedingter Mobilitätsrestriktionen, Hohenstein.

Klepser, Anke, Simone Morlock, Christine Loercher, and Andreas Schenk. 2020. "Functional measurements and mobility restriction (from 3D to 4D scanning)." In *Anthropometry, Apparel Sizing and Design*, 2nd ed., edited by Norsaadah Zakaria and Deepti Gupta, 169–199. Amsterdam: Elsevier.

Klepser, Anke, and Christian Pirch. 2021. "Is this real? Avatar generation for 3D garment simulation." *Journal of Textile and Apparel, Technology and Management* 12: 1–11.

Kozar, T., A. Rudolf, A. Cupar, Simona Jevšnik, and Z. Stjepanovič. 2014. "Designing an adaptive 3D body model suitable for people with limited body abilities." *Journal of Textile Science and Engineering* 4 (5).

Krzywinski, Sybille, and Jana Siegmund. 2017. "3D product development for loose-fitting garments based on parametric human models." 17th World Textile Conference AUTEX 2017, Corfu (Kerkyra) (GR).

Kuijpers, Sandra, Miriam Geelhoed, and Mattijs Crietee. 2021. "Digital fabric roadmap." *International Apparel Federation (IAF).* https://www.iafnet.com/2016_01_22/wp-content/uploads/2021/04/Digital-Fabric-Roadmap-DFR-whitepaper-2021-AMFI-IAF-Modint-powered-by-ClickNL.pdf.

Kuijpers, Sandra, Christiane Luible-Bär, and R. Hugh Gong. 2020. "The measurement of fabric properties for virtual simulation: A critical review." IEEE SA Industry Connections. Industry Connections Report, 1–43. [STDVA24083 978-1-5044-6497-0]. https://standards.ieee.org/content/dam/ieeestandards/standards/web/governance/iccom/3DBP-Measurement_of_fabric_properties.pdf.

Lage, Agne, and Kristina Ancutiene. 2017. "Virtual try-on technologies in the clothing industry. Part 1: Investigation of distance ease between body and garment." *The Journal of the Textile Institute* 108 (10): 1787–1793. doi: 10.1080/00405000.2017.1286701.

Lapkovska, Eva, and Inga Dabolina. 2018. "An investigation on the virtual prototyping validity: Simulation of garment drape." SOCIETY.INTEGRATION.EDUCATION, Proceedings of the International Scientific Conference, Volume IV, May 25–26, 448–458.

Lectra. "Was is Modaris Expert?" Accessed December 29, 2021. www.lectra.com/de/produkte/modaris-expert.

Lee, Eunyoung, and Huiju Park. 2017. "3D virtual fit simulation technology: Strengths and areas of improvement for increased industry adoption." *International Journal of Fashion Design, Technology and Education* 10 (1): 59–70. doi: 10.1080/17543266.2016.1194483.

Lee, Joohyun, Yunja Nam, Ming Hai Cui, Kueng Mi Choi, and Young Lim Choi. 2007. "Fit evaluation of 3D virtual garment." Second International Conference on Usability and Internationalization, Beijing (CN).

Lee, Minjeong, Heesoon Sohn, and Jongjun Kim. 2011. "A study on representation of 3D virtual fabric simulation with drape image analysis II." *Journal of Fashion Business* 15 (3): 97–111.

Leipner, Anja, and S. Krzywinski. 2013. "3D product development based on kinematic human models." 4th International Conference and Exhibition on 3D Body Scanning and Processing Technologies, Long Beach (USA).

Lim, Hosun, and Cynthia L. Istook. 2011. "Drape simulation of three-dimensional virtual garment enabling fabric properties." *Fibers and Polymers* 12 (8): 1077–1082. doi: 10.1007/s12221-011-1077-1.

Liu, Kaixuan, Xianyi Zeng, Pascal Bruniaux, Jianping Wang, Edwin Kamalha, and Xuyuan Tao. 2017. "Fit evaluation of virtual garment try-on by learning from digital pressure data." *Knowledge-Based Systems* 133: 174–182. https://doi.org/10.1016/j.knosys.2017.07.007.

Morlock, Simone. 2015. *Passformgerechte u. bekleidungsphysiologisch optimierte Bekleidungskonstruktion für Männer mit großen Größen unterschiedl. Körpermorphologien.* Hohenstein: HIT.

Morlock, Simone. 2020a. "Passform & Schnitt im Wandel—Mit 3D-Technologie in die Zukunft." *TextilPlus* 9 (10): 13–15.

Morlock, Simone. 2020b. "The transformation of fit and pattern—with 3D towards the future (oral presentation)." Digital Fashion Innovation (DFI) e-Symposium, 28, September 30, online.

Morlock, Simone. 2020c. "Virtual designing and fitting—3D simulation in clothing development." ISPO Masterclass, München (GER).

Morlock, Simone. 2021. "Digital fabric physics (webinar)." Accessed January 4, 2022. www.hohenstein-academy.com/en/e-learning-videos/seminar-detail/show/digital-fabric-parameters-webinar.

Morlock, Simone, Andreas Schenk, Anke Klepser, and Christine Loercher. 2020. "Sizing and fit for plus-size men and women wear." In *Anthropometry, Apparel Sizing and Design*, 2nd ed., edited by Norsaadah Zakaria and Deepti Gupta. Amsterdam: Elsevier.

Morlock, Simone, Ellen Wendt, Elfriede Kirchdörfer, Martin Rupp, Sybille Krzywinski, and Hartmut Rödel. 2009. *Grundsatzuntersuchung zur Konstruktion passformgerechter Bekleidung für Frauen mit starken Figuren.* Hohenstein: Bekleidungsphysiologisches Institut Hohenstein, Technische Universität Dresden.

Naglic, Maja Mahnic, Slavenka Petrak, and Zoran Stjepanovic. 2016. "Analysis of tight fit clothing 3D construction based on parametric and scanned body models." 7th International Conference and Exhibition on 3D Body Scanning and Processing Technologies, Lugano (CH).

Ögülmüs, Esra, Mustafa E. Üreyen, and Cafer Arslan. 2015. "Comparison of real garment design and 3D virtual prototyping." 15th World Textile Conference AUTEX 2015, Bucharest (RO).

Optitex. Accessed December 29, 2021. https://optitex.com/.

Pirch, Christian. 2021. "Digital processing workflow in 3D garment simulation (oral presentation)." Textile and Fashion Innovation Congress (TIFC), online.

Pirch, Christian, Oliver Kausch, Anke Klepser, and Simone Morlock. 2021. "One avatar, two level of detail, one result?: Analyzing the effect of low and high detailed avatars on fitting simulations." 3DBODY.TECH 12th International Conference and Exhibition on 3D Body Scanning and Processing Technologies, Online.

Porterfield, Anne, and Traci A.M. Lamar. 2017. "Examining the effectiveness of virtual fitting with 3D garment simulation." *International Journal of Fashion Design, Technology and Education* 10 (3): 320–330. doi: 10.1080/17543266.2016.1250290.

Power, Jess. 2013. "Fabric objective measurements for commercial 3D virtual garment simulation." *International Journal of Clothing Science and Technology* 25 (6): 423–439. doi: 10.1108/IJCST-12-2012-0080.

Rudolf, A., Z. Stjepanović, and A. Cupar. 2021. "Design of garments using adaptable digital body Models." Tex Teh X—International Conference on Textiles and Connected R&D Domains.

Sayem, Abu Sadat Muhammad. 2017. "Objective analysis of the drape behaviour of virtual shirt, part 2: Technical parameters and findings." *International Journal of Fashion Design, Technology and Education* 10 (2): 180–189.

Sayem, Abu Sadat Muhammad. 2019. "Virtual fashion ID: A reality check." IFFTI Conference, April 8–11, Manchester Fashion Institute, Manchester (GB).

Song, Hwa Kyung, and Susan P. Ashdown. 2015. "Investigation of the validity of 3-D virtual fitting for pants." *Clothing and Textiles Research Journal* 33 (4): 314–330. doi: 10.1177/0887302x15592472.

VitalMechanics. Accessed September 10, 2022. www.vitalmechanics.com/#products.

Volino, Pascal, Frederic Cordier, and Nadia Magnenat-Thalmann. 2005. "From early virtual garment simulation to interactive fashion design." *Computer-Aided Design* 37 (6): 593–608. https://doi.org/10.1016/j.cad.2004.09.003.

Wu, Yen Yu et al. 2011. "An investigation on the validity of 3D clothing simulation for garment fit evaluation." International conference on Innovative Methods in Product Design, Venedig, Italy.

Zangue, Flora, Christian Pirch, and Anke Klepser. 2022. "Evaluation of garment simulations within an analytical fi^t test." 14th Joint International Conference CLOTECH, September 5–8.

Zangue, Flora, Christian Pirch, Anke Klepser, and Simone Morlock. 2020. "Virtual fit vs. physical fit: How well does 3D simulation represent the physical reality." 3DBODY.TECH 2020 11th Int. Conference and Exhibition on 3D Body Scanning and Processing Technologies, Online/Lugano (CH).

Zhang, D., and S. Krzywinski. 2019. "Development of a kinematic human model for clothing and high performance garments." 3DBODY.TECH 10th International Conference and Exhibition on 3D Body Scanning and Processing Technologies, Lugano (CH).

4 Virtual Fit of Bodices Constructed Following Contemporary Methods

Md. Mazharul Islam and Abu Sadat Muhammad Sayem

CONTENTS

4.1 INTRODUCTION

A key step in the clothing development process is pattern cutting that works as an aid in cutting fabric prior to garment manufacture. In fashion dictionary, patterns represent the geometrical drawing of outlines of the components that will make a full garment when assembled. Traditionally, a tailor or a pattern technician converts the initial illustrations by designers into technical designs in the form of pattern pieces by incorporating the body size measurements, ease and seam allowances and other information necessary for the subsequent production process. The basic approaches that are traditionally followed in the industry for producing pattern pieces are 3D draping (also known as the "*haute couture*" process) and pattern drafting.

The process of 3D draping or modelling starts with the production of "toile", an intermediate garment, by draping, moulding, cutting and pinning cheap fabric like calico on to a body form, often a size-specific mannequin directly, but sometimes also on to a human body. The key components of the toile are then cut off the body

DOI: 10.1201/9781003264958-6

form to produce individual flat pieces, which act as pattern pieces for that style. Additional measurement allowances for wear comfort and sewing seams are then added while transferring them into paper patterns. This is fully a manual and time-consuming process, but when it comes to a good fit and flattering drape, often as per the requirement of specific customers, this process is very effective. This inherently incorporates 3D anthropometry during toile making.

The engineered version of pattern cutting that relies on geometrical drawing technique and makes use of average size measurements to produce pattern pieces is known as pattern drafting. The process is commonly practised in the industry targeting the mass production of apparel made for standard average groups of population. Block patterns represent a set of pattern pieces for a specific style of garment targeting to fit an average figure (Aldrich 1990). Commonly followed block pattern construction methods that are taught in academic set up to train fashion designers and clothing technicians are the methods by Aldrich (2008), Abling and Maggio (2009), Beazley and Bond (2003), Bunka (2009), Burgo (2004), Hundt (2016), Joseph-Armstrong (2010) and Shoben and Ward (1987). These pattern construction methods differ from each other and are not perfect and absolute. Although they are known to the industry for years, it requires the experience and expertise of pattern cutters to adapt them to achieve a desired fit. This chapter aims to facilitate virtual fashion designers and pattern technicians with evidence of virtual fit comparison among the said methods. Although virtual try-on technique has been applied to study distance ease between body and garment (Lage and Ancutiene 2017; Liu et al. 2019), virtual fits of 3D simulation of block patterns by different pattern construction methods have not been studied before.

4.2 SUPPORTING LITERATURE

4.2.1 Ease in Pattern

The linear distance between a wearer's skin to the inner layer of clothing item being worn is known as "ease". Based on the purpose of use, ease can be either functional ease or aesthetic (or styling). Functional ease is added to pattern pieces to ensure wearers' comfort during body expansion while breathing in and out and dynamic movement of limbs. It is known that the girth measurements of chest/bust, neck and waist vary during inhalation and exhalation (Moll and Wright 1972; Mckinnon and Istook 2002). Skin dimensions significantly change during dynamic actions (Chi and Kennon 2006; Ziegert and Keil 1988), which necessitates functional ease in patterns. The aesthetic or style ease is the added measurements to pattern pieces required to reproduce any specific silhouette. Usually, pattern technicians add the required level of ease reflecting the designers' concept and need of the target consumers. However, there is no established formula or guideline of determining required ease of any specific style, rather than depending on the experience of pattern cutters (Petrova and Ashdown 2008).

The contemporary pattern construction methods recommend different values of ease in the same location of patterns and in most cases do not provide justification against those recommendations (Aldrich 2008; Abling and Maggio 2009; Beazley and Bond 2003; Bunka 2009; Burgo 2004; Hundt 2016; Joseph-Armstrong 2010; Shoben and Ward 1987). An objective of this study is to compare the ease values used in the key pattern locations in bodice blocks from these pattern construction methods.

4.2.2 Computer-Aided Design (CAD) Technology for Fashion

CAD techniques have become popular in pattern drafting since the 1980s. It is now a common practice in the industry to use CAD systems to draft patterns geometrically. Several 2D CAD systems are now available on the market. These systems also accept input from a "digitiser" to generate digital version of any physical pattern piece. 3D techniques to produce virtual clothing started to evolve in the 1990s (Okabe et al. 1992; Volino et al. 1996; Luo and Yue 2005). Today several 3D systems that can drape 2D digital pattern pieces onto virtual avatars are available for industrial application (Sayem 2017a, 2017b; Sayem and Bednall 2017). These software packages come with a library of digital avatars and fabrics and make use of computer graphics tools to simulate digital patterns into virtual garments. They also provide some technical tools such as a tension map, stress map, ease map, etc. to facilitate the virtual check of clothing fit onto the virtual avatar.

Another approach to generate patterns is to extract flat pattern directly from 3D design using surface flattening tools, and a notable amount of research work has been conducted on this (McCartney et al. 2000; Kim and Kang 2002; Petrak et al. 2006; Fang et al. 2008). Although the technique is very promising, the software technology is not yet advanced enough to extract pattern pieces for both intimate wear and outerwear that would attract industrial attention.

This work utilises the traditional approach of 2D to 3D drape simulation in producing the virtual garments from digital 2D block patterns.

4.3 METHODOLOGY

Bodice blocks from the contemporary pattern construction methods by Aldrich (2008), Abling and Maggio (2009), Beazley and Bond (2003), Bunka (2009), Burgo (2004), Hundt (2016), Joseph-Armstrong (2010), Shoben and Ward (1987) are chosen for this work to produce eight sets of digital patterns. The characteristics of these 2D digital pattern blocks are compared among one another. Later the digital block patterns are simulated into 3D virtual garments, and their virtual drape parameters are compared both subjectively and objectively. The step-by-step workflow is elaborated in the following subsections.

4.3.1 Pattern Drafting

Digital patterns for bodice blocks using the chosen methods are constructed using an industry standard 2D CAD pattern drafting system following the size 10 female body measurements presented in Joseph-Armstrong (2010). The measurement chart from Joseph-Armstrong (2010) was chosen because it provides the most compressive body measurements that meet the needs of all other pattern construction methods, as can be seen in Table 4.1. As shown in Figure 4.1, the patterns are first geometrically

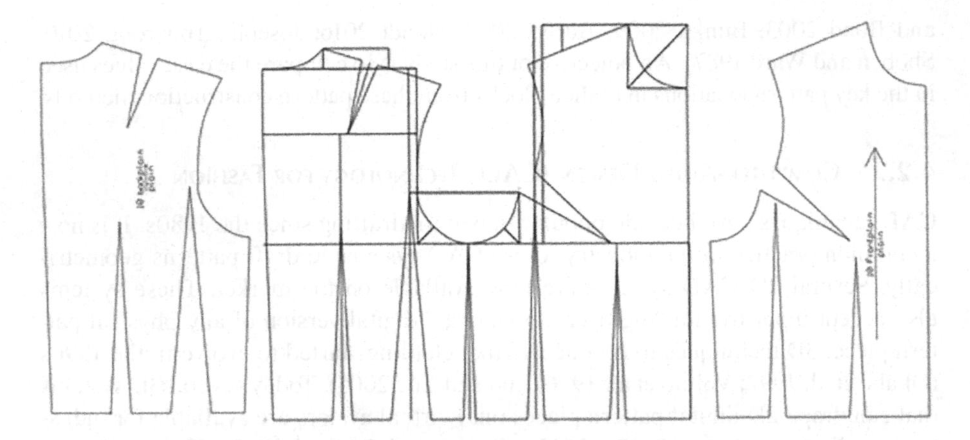

FIGURE 4.1 Example of bodice pattern construction (far left and right shows the back and front parts and middle section represents the drafts).

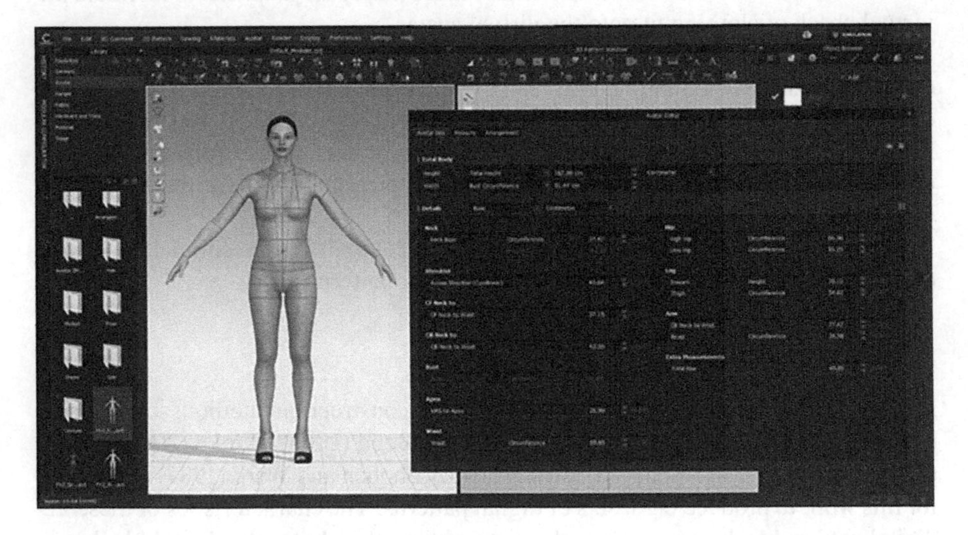

FIGURE 4.2 Screenshot of 3D avatar window.

drafted on the 2D CAD window of the software system and traced off as clean patterns which were stored in .dxf data format for export to the 3D CAD system.

4.3.2 Avatar Morphing

An available virtual avatar named "feifei v2" from the avatar library of the industry grade software system CLO 3D was chosen. The key measurements of this avatar were modified according to Table 4.1 using the morphing tools in the avatar editor window (see Figure 4.2) to make it as a representative body shape for the drafted bodice blocks.

4.3.3 3D Stitching, Simulation and Fit Mapping

All pattern sets are individually placed at the front and back sides of the avatar, and the front and back patterns are virtually stitched at shoulder seam and side seams (see Figure 4.3). The edges of the darts are also stitched together to leave no gap on the virtual garments. An available woven fabric of "muslin 18 × 20.zfav" representing 100% cotton, weighing 103.03 g/m^2, with a thickness of 0.3 mm is selected from the material library of Clo3D system to simulate the virtual garment from each of the eight pattern sets. For each virtual garment, the drape and fit quality are analysed using the virtual parameters like stress map, strain map, fit map and pressure points and opaque surface view for wrinkles (see Figure 4.4).

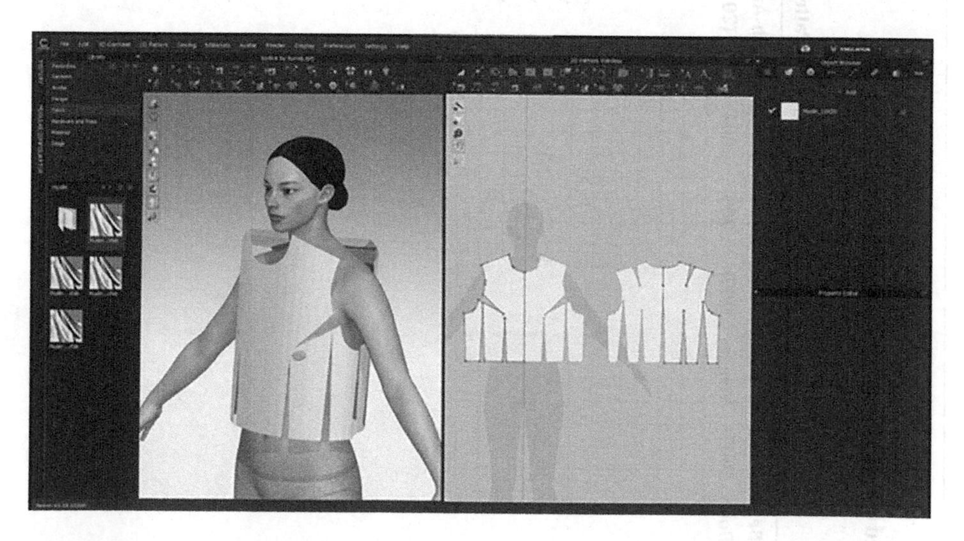

FIGURE 4.3 2D pattern placement of 3D avatar and virtual stitching.

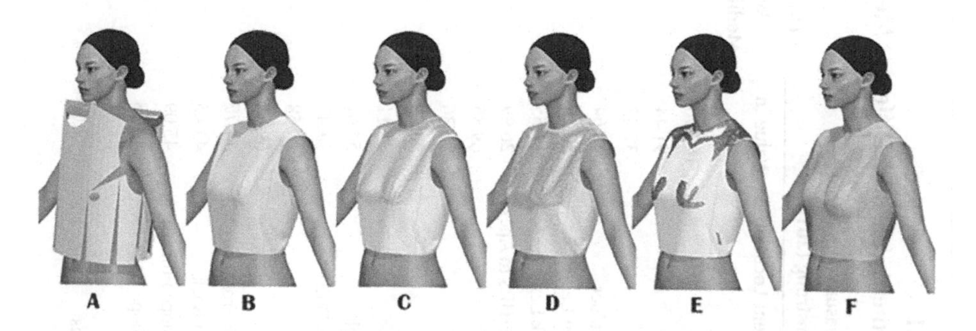

FIGURE 4.4 Different steps in virtual simulation and fit mapping: (A) virtual stitching, (B) 3D drape after simulation, (C) stress map, (D) strain map, (E) fit map and pressure points and (F) opaque surface.

TABLE 4.1
Measurements Used by Different Methods and for Avatar Modification

Size 10 Measurements from Joseph-Armstrong (2010)		Used for Avatar Modification	Measurements Used in Pattern Drafting							
Measurement Locations	Value in cm		Beazley and Bond (2003)	Abling and Maggio (2009)	Bunka (2009)	Burgo (2004)	Joseph-Armstrong (2010)	Shoben and Ward (1987)	Aldrich (2008)	Hundt (2016)
Height	167.96	√				√				
Bust	91.44	√	√	√	√	√		√	√	√
Neck Base	37.47	√	√					√	√	√
Across Shoulder	40.64	√		√		√				
Centre Front Length	37.15	√	√	√			√			
Centre Back Length	42.55	√	√	√	√		√	√	√	√
Bust Depth (HPS to Apex)	26.99	√	√	√		√	√	√		√
Waist	69.85	√	√	√	√	√		√	√	√
Abdomen	86.36	√								
Hip	95.25	√								
Inseam	78.11	√								
Upper Thigh	54.61	√								
CB Neck to Wrist	77.47	√								
Bicep	28.58	√								
Total Rise	69.85	√								
Full Length Front	44.77			√		√	√			√
Full Length Back	44.45			√		√	√			
Shoulder Slope Front	45.09			√			√			
Shoulder Slope Back	43.18			√			√			
Strap	45.09						√			
Bust Radius	7.62									

Bust Span (Apex to Apex)	9.84	√	√	√		√	√			√
Side Length	21.27			√			√			
Back Neck	7.62						√			
Shoulder Length	13.34		√	√			√	√	√	√
1/2 across Shoulder Front	19.69						√			
1/2 across Shoulder Back	20.32						√			
Half across Chest	17.15		√	√			√		√	
Half across Back	17.78		√	√		√	√	√	√	√
Bust Arc	25.4						√			
Back Arc	21.91						√			
Waist Arc Front	17.78						√			
Waist Arc Back	16.83						√			
Dart Placement	8.26						√			
Arm Scye Depth	21		√					√	√	√
Armhole Width	11.9							√		
Armhole Girth	44.6							√		
Dart	7								√	
Total	**38**	**16**	**11**	**15**	**3**	**9**	**21**	**10**	**9**	**10**

4.4 RESULTS AND DISCUSSIONS

4.4.1 Comparison of 2D Patterns

As can be seen in Table 4.1, Bunka (2009) uses only three measurements to produce the front and back bodice patterns, whereas the Joseph-Armstrong (2010) method requires 21 measurements for producing the pattern set. Table 4.2 and Figure 4.5 present the shape and dimensional comparisons of 2D pattern blocks of a bodice constructed based on the same body measurements using eight different methods. It is clear that each method produces different pattern shapes and dimension from others. When the lengths from high point shoulder (HPS) to bottom are compared, Burgu's (2004) method produces the shortest front bodice, and Bunka's (2009) produces the tallest front bodice. However, in the case of the back bodice, the pattern by Hundt (2016) is the shortest, and the pattern by Abling and Maggio (2009) is the longest. Joseph-Armstrong (2010) produces the widest front bodice, whereas the front bodice by Abling and Maggio (2009) is the narrowest. In the case of the back bodice, Shoben and Ward's (1987) is found to be the widest, and Hundt's (2016) is the narrowest.

The front bodice by Joseph-Armstrong (2010) features only one dart but the same pattern by Bunka (2009) includes three darts. All other front bodices include two darts each. In the case of the back bodice, Bunka (2009) again includes three darts, Burgo (2004) and Shoben and Ward (1987) include just one dart and all others include two darts each. This is also reflected on the level of ease incorporated into the pattern pieces (see Table 4.3). At the bust girth, Bunka (2009) incorporates the highest ease, and Beazley and Bond (2003) does the least. At the waist girth, Burgo (2004) has the maximum ease, whereas Abling and Maggio (2009) add the minimum.

4.4.2 Comparison of Stress and Strain Maps

Figures 4.6 and 4.7 provide the stress maps and strain maps respectively of the 3D virtual garments simulated from different bodice blocks constructed for this study. The 3D CAD system we used indicates the pressure acting on the virtual surface in Kilo Pascal (kPa) and presents the stress map with colour codes where blue indicates zero stress (0.00kPa) while red indicates the most stress (100kPa). On screen all stress maps appeared in light blue shade on the stressed areas, although presented as a black and white image in the Figure 4.6. This indicates the acting stress on all virtual garments are very low, which is also evident in Figure 4.6. This is also similar in case of strain map that visualises the stretched areas on the virtual garment (see Figure 4.7). The stress map is also represented with colour codes where blue indicates 100% of the distortion rate (no distortion), and red indicates 120% of the distortion rate. The strain maps of all bodices are mostly blue except the waist region of the bodice from Joseph-Armstrong (2010) and sides of Aldrich (2008), although it is represented as a black and white image in the Figure 4.7. This indicates a low level of strain on the digital fabrics.

Only visual assessment of the stress and strain map does not provide any reliable clue as to the virtual fit of the clothing (Sayem 2017a, 2017b; Sayem and Bednall 2017). The appearances of most of the stress and strain maps in Figures 4.6 and 4.7 are pretty similar. Therefore, the numerical values of stress (KPa) and strain (%) are

TABLE 4.2
2D Pattern Shapes and Dimensions

Methods	Front Bodice			Back Bodice		
	2D Pattern	Length (cm)	Width (cm)	2D Pattern	Length (cm)	Width (cm)
Beazley & Bond (2003)	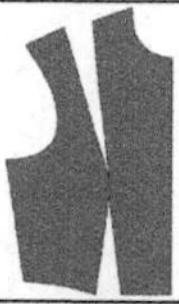	45.04	26.11	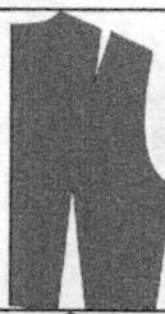	45.05	24.11
Abling & Maggio (2009)	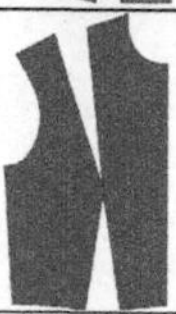	45.61	25.08	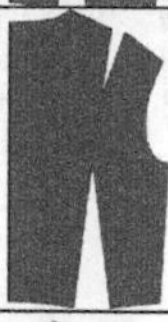	45.29	23.82
Bunka (2009)	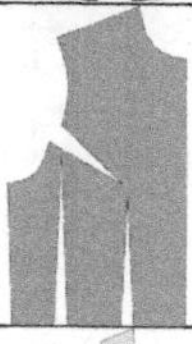	47.82	26.69	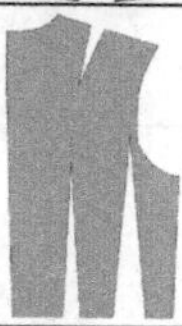	45.02	25.03
Burgo (2004)	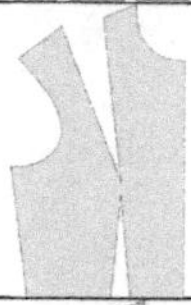	44.77	26.4	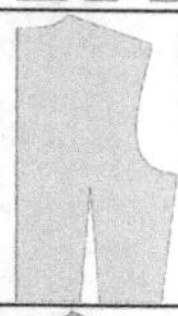	44.45	24.4
Joseph-Armstrong (2010)	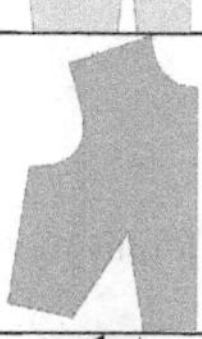	45.57	29.2	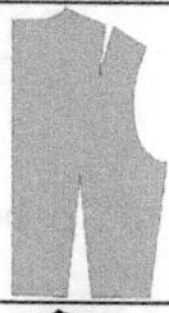	44.77	23.82
Shoben & Ward (1987)	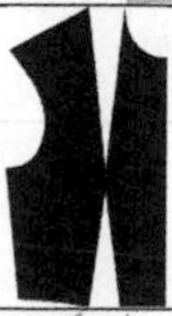	46.38	25.19	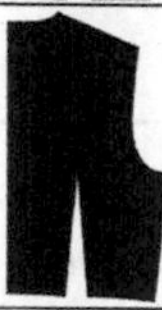	45.25	25.53
Aldrich (2008)	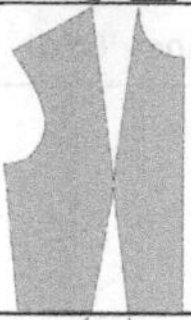	45.05	26.54	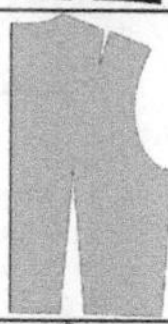	44.62	24.18
Hundt (2016)	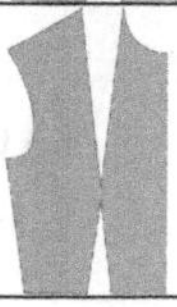	45.27	25.58	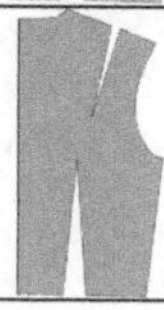	44.39	22.64

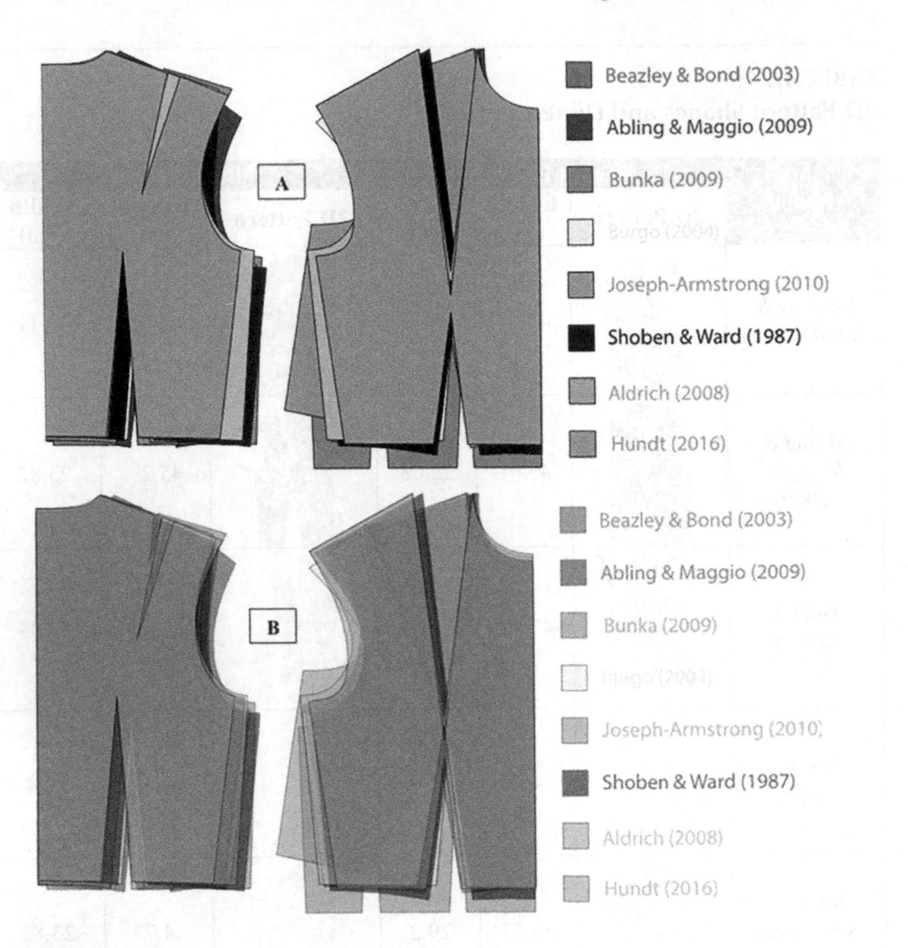

FIGURE 4.5 Comparison of 2D bodice block patterns by different methods: (A) at 100% opacity, (B) at 50% opacity.

TABLE 4.3
Ease Comparison at Key Girth Measurements

Methods	Ease Allowance (cm)	
	Bust Girth	Waist Girth
Beazley and Bond (2003)	6.00	4.00
Abling and Maggio (2009)	6.35	2.54
Bunka (2009)	12.00	6.00
Burgo (2004)	10.16	10.16
Joseph-Armstrong (2010)	N/A (used bust arc + 0.64 cm)	N/A (used back arc + 1.91)
Shoben and Ward (1987)	10.00	7.00
Aldrich (2008)	10.00	6.00
Hundt (2016)	10.00	4.00

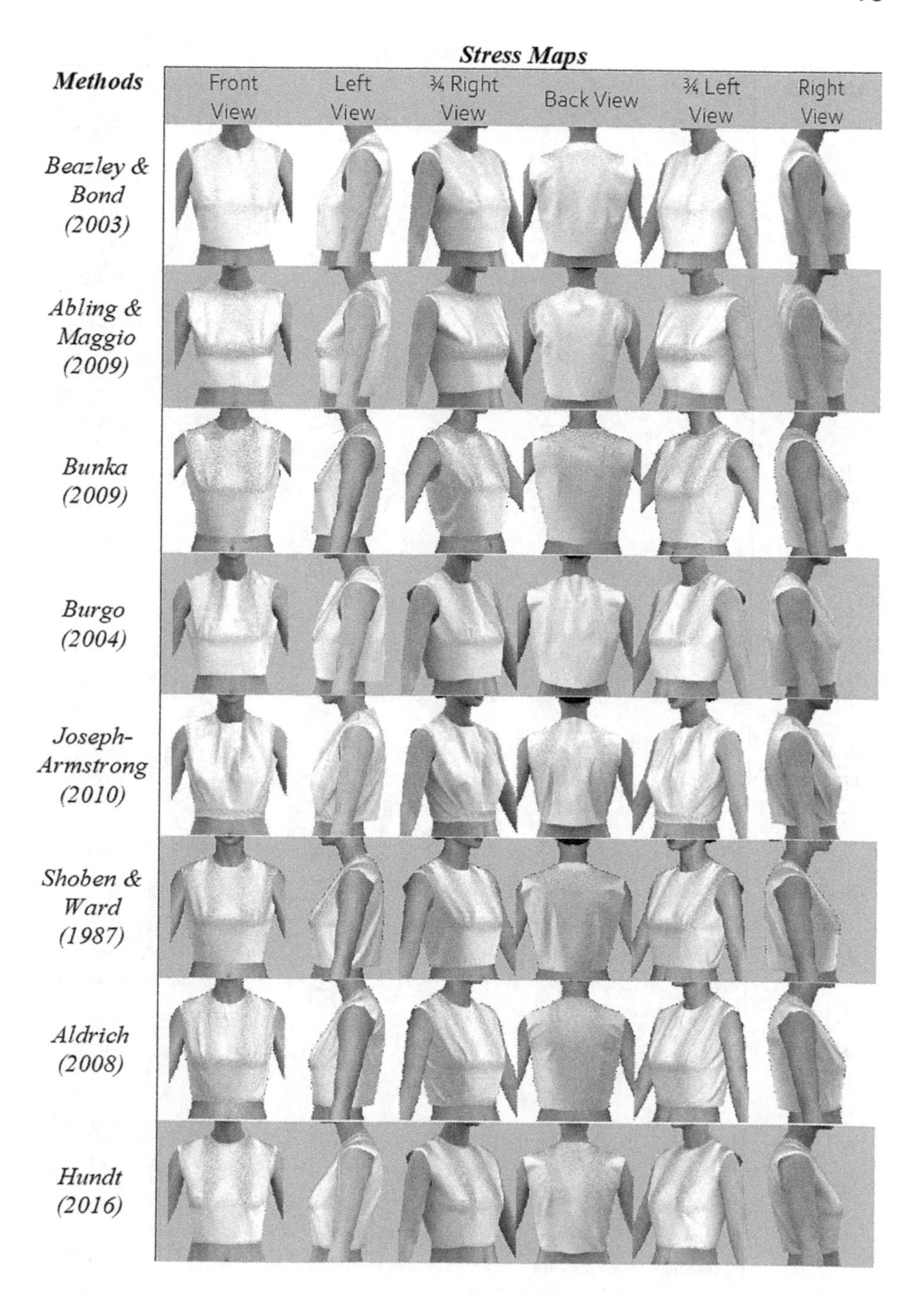

FIGURE 4.6 Comparison of stress maps of different bodice blocks.

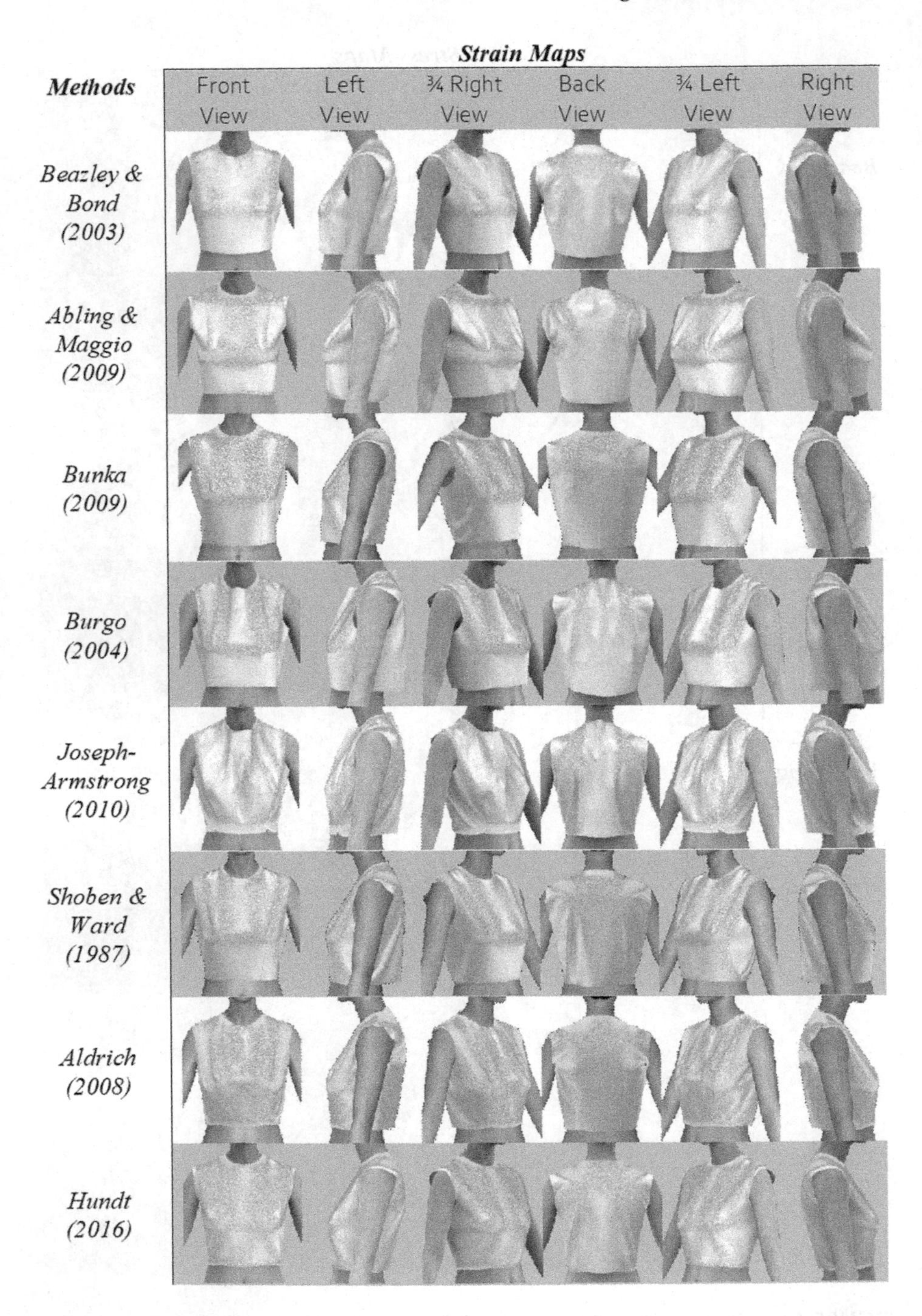

FIGURE 4.7 Comparison of strain maps of different bodice blocks.

considered to understand the comparative picture of the virtual drape produced by the blocks from all selected methods (see Table 4.4, Figure 4.8 and 4.9). When the maximum stress at the bust region is considered, the bodices by Abling and Maggio (2009) exhibit the highest value of 6.9 KPa, and Burgo's (2004) represents the lowest stress

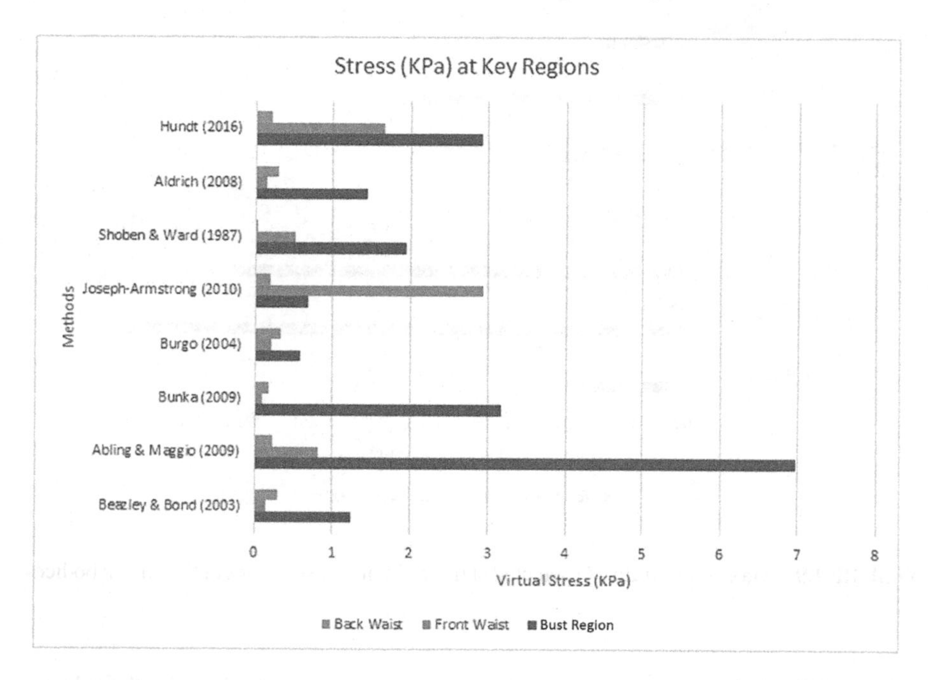

FIGURE 4.8 Maximum stress (KPa) detected at the waist and bust regions of the virtual bodices.

TABLE 4.4
Stress and Strain Values in Key Regions of Pattern

Methods of Bodice Block	Stress Value (KPa)			Strain Value (%)		
	Bust Region	Front Waist	Back Waist	Bust Region	Front Waist	Back Waist
Beazley and Bond (2003)	1.23	0.15	0.30	100.81	100.08	100.14
Abling and Maggio (2009)	6.97	0.82	0.23	104.57	100.54	100.11
Bunka (2009)	3.18	0.09	0.18	103.57	100.04	100.09
Burgo (2004)	0.59	0.22	0.33	100.63	100.14	100.15
Joseph-Armstrong (2010)	0.69	2.93	0.20	100.98	101.92	100.10
Shoben and Ward (1987)	1.94	0.52	0.02	102.10	100.43	100.05
Aldrich (2008)	1.44	0.15	0.30	100.68	100.10	100.14
Hundt (2016)	2.92	1.65	0.21	101.91	101.08	100.10

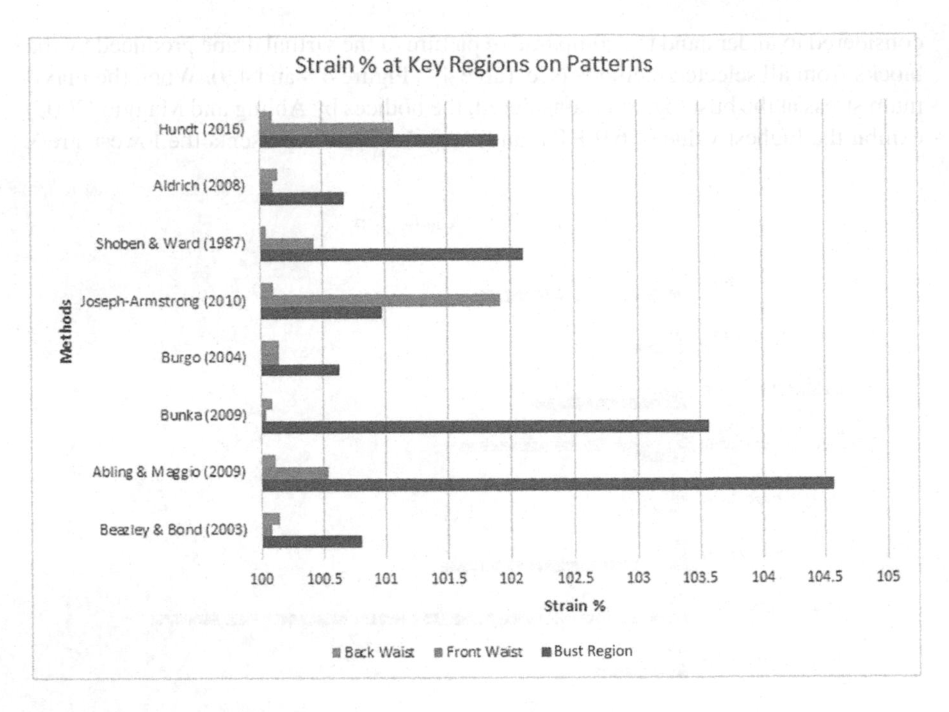

FIGURE 4.9 Maximum strain (%) detected at the waist and bust regions of the virtual bodices.

of 0.59 KPa. At bust region, the order of maximum stress in bodices by the selected methods are: Abling and Maggio (2009), Bunka (2009) KPa, Hundt (2016), Shoben and Ward (1987), Aldrich (2008), Beazley and Bond (2003), Joseph-Armstrong (2010), Burgo (2004) KPa. At front waist of the bodices, Joseph-Armstrong's (2010) method indicates the highest stress value of 2.93 KPa, and Bunka's (2009) shows the lowest value of 0.09 KPa. The order of stress at the front waist is as follows: Joseph-Armstrong (2010), Hundt (2016), Abling and Maggio (2009), Shoben and Ward (1987), Burgo (2004), Beazley and Bond (2003), Aldrich (2008) and Bunka (2009). At the back waist, the acting stress is found to be significantly low with the Burgo (2004) bodice having 0.33 KPa only. This is because the fit at the back waist for most of the bodices is very poor, which is further discussed in Section 4.4.3. However, the order of stress at the back waist of the bodices by different methods is as follows: Burgo (2004), Beazley and Bond (2003), Aldrich (2008), Abling and Maggio (2009), Hundt (2016), Joseph-Armstrong (2010), Bunka (2009) and Shoben and Ward (1987).

When strain values are analysed, the bodice by Abling and Maggio shows the highest strain of 104.57%, and the bodice by Burgo (2004) shows the least strain of 100.63% at the bust region. In this case, the order of strain % is as follows: Abling and Maggio (2009), Bunka (2009), Hundt (2016), Shoben and Ward (1987), Aldrich (2008), Beazley and Bond (2003), Joseph-Armstrong (2010) and Burgo (2004), which shows the same pattern as the stress values.

At the front side of the waist, the maximum strain varies between 100.04% and 101.92%, which can be considered as very minimum. The order of this variation from the highest to the lowest is as follows: Joseph-Armstrong (2010), Hundt (2016), Abling and Maggio (2009), Shoben and Ward (1987), Burgo (2004), Beazley and Bond (2003); Aldrich (2008) and Bunka (2009). The strain at the back waist is very low and indicates a gap between body and garments.

4.4.3 Comparison of Fit Maps and Surface Views

Figures 4.10 and 4.11 compare the fit maps and opaque surface views of the digital bodices. The fit maps in the 3D system in use shows how many sections of the garment have reached the maximum strain limit of the fabric, and it is for the areas in which fabric and avatar are in contact. This is a visual indication of the tightness and presented with colour codes, such as red is equivalent to too tight (Can't Wear) and yellow means minimum tight when presented on a coloured screen. Together the fit maps and the opaque surface views are useful in identifying wrinkles, folds and the areas where fabric is not in contact with the skin. However, they provide conflicting impressions when compared against the stress and strain maps. According to Figures 4.6 and 4.8, the bust area of the bodice by Abling and Maggio (2009) is most stressed among all others, but the fit map in Figure 4.10 shows the densest pressure points on the bust area of the bodice by Aldrich (2008).

When the fit maps and opaque surface views are jointly analysed, the bodices by Aldrich (2008), Shoben and Ward (1987) and Joseph-Armstrong (2010) show noticeable wrinkles on the front sides. At the back waist, all the bodices except that of Hundt (2016) show very poor fit, having big gaps between fabric and avatar skin (see left and right views in Figures 4.10 and 4.11).

4.5 CONCLUSION

It is apparent from the analysis that the contemporary pattern cutting methods differ to each other in respect of the measurement locations they use to construct bodice block, ease in key regions and 2D and 3D shapes. As a result, they exhibit different virtual fits of resultant virtual bodices when compared both objectively and subjectively. The stress at the bust and front waist regions differs significantly among the different bodices, but at a range of 0.59 ~ 6.9 KPa and 0.09 ~ 2.93 KPa, respectively, and the strain at bust and front waist differ at a range of 100.63 ~ 104.57% and waist 100.04% ~ 101.92%, respectively. However, it is unclear how these ranges of stress and strain variations will impact the fit and user comfort of physical prototypes. This demands further work with physical prototypes of corresponding virtual prototypes and wearer trials with live models to establish the true accuracy rating of the methods. This chapter has presented the virtual fit analysis of bodice blocks produced by contemporary methods, which was not presented before. As virtual fashion prototyping is gaining popularity in the industry, these findings will provide practical clues to the digital designer and professional about the selection of the right method of pattern cutting for industrial application.

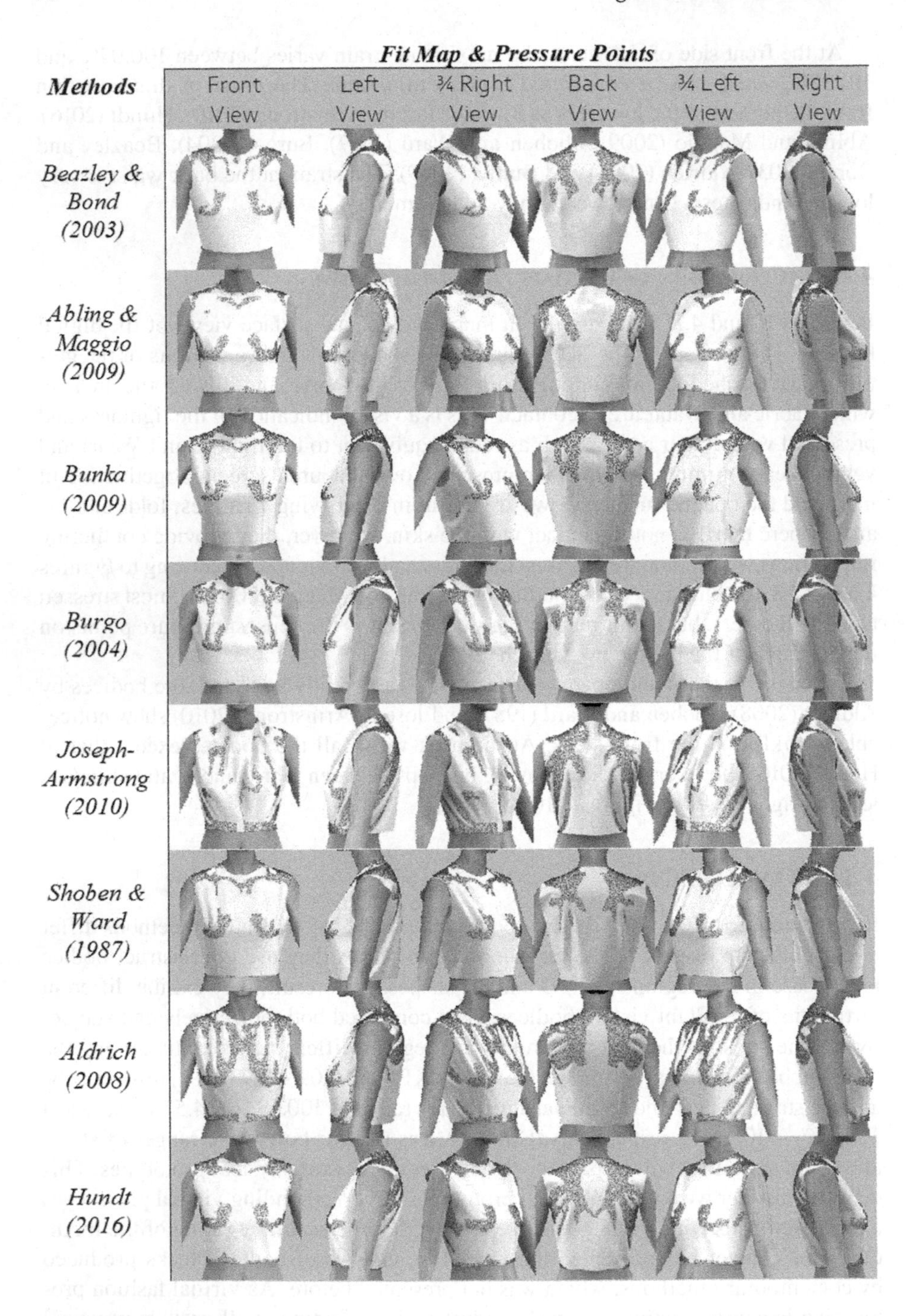

FIGURE 4.10 Comparison of fit maps and pressure points of different bodice blocks.

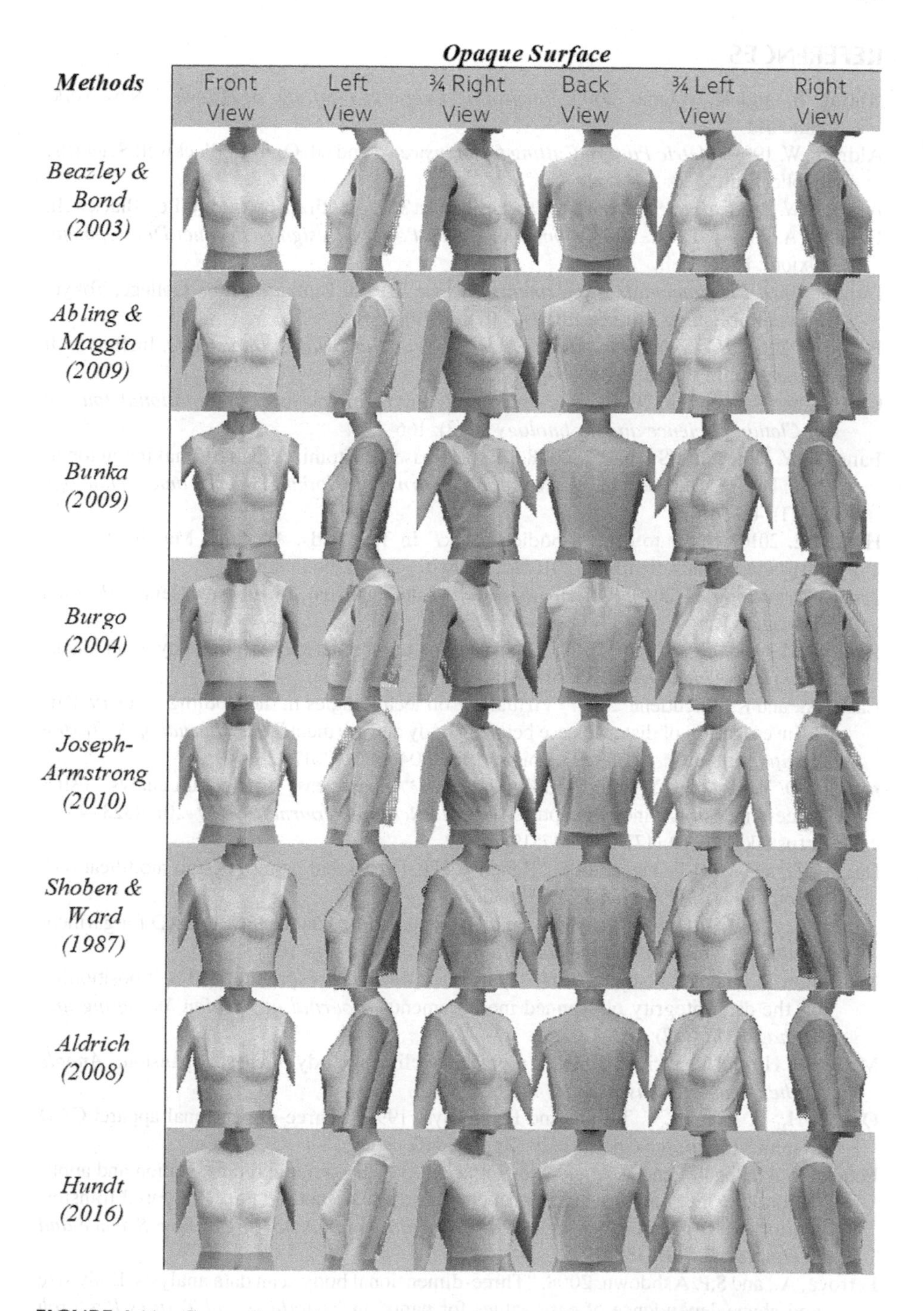

FIGURE 4.11 Comparison of opaque surface views of different bodice blocks.

REFERENCES

Abling, B., and K. Maggio. 2009. *Integrating Draping, Drafting, & Drawing.* New York: Fairchild.

Aldrich, W. 1990. *Metric Pattern Cutting for Menswear*, 2nd ed. Oxford: Blackwell Scientific Publications.

Aldrich, W. 2008. *Metric Pattern Cutting for Women's Wear*, 5th ed. Oxford: Wiley-Blackwell.

Beazley, A., and T. Bond. 2003. *Computer-Aided Pattern Design & Product Development.* Oxford: Blackwell.

Bunka. 2009. *Fundamentals of Garments Design*, 1st ed. Bunka Fashion College, Tokyo: Bunka Publishing Bureau, ISBN-13:978-4579112388.

Burgo, F. 2004. *IL Modellismo—Tecnica del Modello Sartoriale e Industriale.* Instituto di Milan: Moda Burgo. ISBN 88–900101–50

Chi, L., and R. Kennon. 2006. "Body scanning of dynamic posture." *International Journal of Clothing Science and Technology* 18 (3): 166–178.

Fang, J., Y. Ding, and S. Huang. 2008. "Expert-based customized pattern-making automation: Part II. Dart design." *International Journal of Clothing Science and Technology* 20 (1): 41–56.

Hundt, E. 2016. "How to draft a bodice block." In The folds. Accessed March 14, 2021. https://inthefolds.com/blog/2016/2/22/how-to-draft-a-bodice-block.

Joseph-Armstrong, H. 2010. *Patternmaking for Fashion Design*, 5th ed. New Hersey: Pearson Education Inc.

Kim, S.M., and T.J. Kang. 2002. "Garment pattern generation from body scan data." *Computer-Aided Design* 35: 611–618.

Lage, A., and K. Ancutiene. 2017. "Virtual try-on technologies in the clothing industry. Part 1: Investigation of distance ease between body and garment." *The Journal of the Textile Institute* 108 (10): 1787–1793. doi: 10.1080/00405000.2017.1286701

Liu, Z., Q. He, F. Zou, Y. Ding, and B. Xu. 2019. "Apparel ease distribution analysis using three-dimensional motion capture." *Textile Research Journal* 89 (19–20): 4323–4335. https://doi.org/10.1177/0040517519832842

Luo, G.Z., and M.M.F. Yuen. 2005. "Reactive 2D/3D garment pattern design modification." *Computer-Aided Design* 37: 623–630.

McCartney, J., B.K. Hinds, B.L. Seow, and D. Gong. 2000. "Dedicated 3D CAD for garment modelling." *Journal of Materials Processing Technology* 107: 31–36.

Mckinnon, L., and C.L. Istook. 2002. "The effect of subject respiration and foot positioning on the data integrity of scanned measurements." *Journal of Fashion Marketing and Management* 6 (2): 103–121.

Moll, J.M.H., and V. Wright. 1972. "An objective clinical study of chest expansion." *Annals of the Rheumatic Diseases* 31: 1–8.

Okabe, H., H. Imaoka, T. Tomih, and H. Niwaya. 1992. "Three-dimensional apparel CAD system." *Computer Graphics* 26 (2): 105–110.

Petrak, S., D. Rogale, and V. Mandekić-Botteri. 2006. "Systematic representation and application of a 3D computer-aided garment construction method Part II: Spatial transformation of 3D garment cut segments." *International Journal of Clothing Science and Technology* 18 (3): 188–199.

Petrova, A., and S.P. Ashdown. 2008. "Three-dimensional body scan data analysis: Body size and shape dependence of ease values for pants' fit." *Clothing and Textiles Research Journal* 26 (3): 227–252.

Sayem, A.S.M. 2017a. "Objective analysis of the drape behaviour of virtual shirt, part 2: Technical parameters and findings." *International Journal of Fashion Design,*

Technology and Education 10 (2): 180–189. http://dx.doi.org/10.1080/17543266.2016.1223810.

Sayem, A.S.M. 2017b. "Objective analysis of the drape behaviour of virtual shirt, part 1: Avatar morphing and virtual stitching." *International Journal of Fashion Design, Technology and Education* 10 (2): 158–169. http://dx.doi.org/10.1080/17543266.2016.1223354.

Sayem, A.S.M., and A. Bednall. 2017. "A novel approach to fit analysis of virtual fashion clothing." 19th edition of the International Foundation of Fashion Technology Institutes conference (iffti 2017), IFFTI 2017, The Amsterdam Fashion Institute (AMFI), Amsterdam, 29/3/2017–30/3/2017.

Shoben, M.M., and J.P. Ward. 1987. *Pattern Cutting and Making Up: The Professional Approach*, Vol 1, revised ed. Oxfordshire: Routledge.

Volino, P., N.M. Thalmann, J. Shen, and D. Thalmann. 1996. "An evolving system for simulating clothes on virtual actors." *IEEE Computer Graphics and Applications* 16 (5): 42–51.

Ziegert, B., and G. Keil. 1988. "Stretch fabric interaction with action wearable: Defining a body contouring pattern system." *Clothing Research Journal* 6: 54–64.

5 Scan2Weave *Connecting Digital Anthropometry with 3D Weaving Technology*

Yuyuan Shi, Lindsey Waterton Taylor, Vien Cheung and Abu Sadat Muhammad Sayem

CONTENTS

5.1 INTRODUCTION

Anthropometry is an indispensable foundation in the process of clothing design and production. Dressmakers formerly obtained primary clothing data through manual measuring, such as chest circumference, waist circumference and hip circumference. Three-dimensional (3D) scanning technology, originated at the end of the last century, has developed rapidly in decades (Daanen and Ter Haar 2013). The 3D scanners that are primarily applied in the clothing industry can be identified as laser scanning systems (Daanen and Van De Water 1998; Robinette and Daanen 2006), structured light systems (Mochimaru and Kouchi 2011), multi-view camera systems (van Iersel et al. 2009) and millimetre wave systems for 3D static and 4D dynamic measurements. Comparing with traditional measurement, the 3D scanning equipment can obtain thousands of more accurate data in seconds. This dramatically saves time and manpower costs. Aligned with equipment and software development, altered kinds of digital software are additionally developed to optimise, gather and analyse the scanning database.

3D digital platforms applied in the clothing industry are divided into two categories. One set is available to design and create garments directly in a 3D virtual

DOI: 10.1201/9781003264958-7

environment consistent with personal priorities, such as TPC Parametric Pattern Generator (TPC, HK); the other is to simulate 3D design versions based on the imported 2D information, for instance, Accumark Vstistcher™ (Gerber) and Modaris 3D FIT (Lectra) (Sayem et al. 2010). In addition to specially aimed clothing software, the multi-faced engineering software systems, such as Geomegic package (3D Systems, Inc.) and Design Concept 3D TechTex (Lectra), are introduced in the clothing field (Sayem et al. 2010; Sayem 2017) to allow the development of potential cross-discipline technology. Applying 3D scanning devices and software in clothing anthropometry is beneficial to solve the problem of fitness in-depth and provide the basis for product development of close-fitting functional clothing—sportswear or sports bras.

Computer-aided design and manufacturing (CAD/CAM) applied in different clothing design stages and production significantly promotes routine jobs (Sayem et al. 2010; Dāboliņa et al. 2018). The 2D CAD/CAM system with an affordable price is broadly applied in the clothing pattern construction, fabric cutting and spreading process to save time, increase efficiency and reduce labour cost. The pattern construction with the aid of the 2D digital platform realises rapid standardised parametric drawing, which facilitates collaborations and communications among designers, manufacturers and retailers remotely or globally (Kennedy 2015). Besides that, the aimed-clothing CAD system containing the various grading rules obtained manually or digitally (Kennedy 2015) is appropriate for the mass production of garments. The automatic fabric pattern process also boosters pre-preparation for the final sewing process.

Cut-and-sew is the primary manufacturing process for daily life clothing in the apparel industry (Kunz and Glock 2004). However, this conventional technique is not applicable for functional or protective garments (Colovic 2015). Discontinuity of fibre-yarn due to the cutting of textiles reduces the level of reinforcement and ultimate protection (Chen and Yang 2010a, 2010b). Seams derived from this traditional manufacture process are not adequate for sportswear, as the direct contact between textiles and skin is vital in the tactile comfort of clothing (Bartels 2006). This irritating discomfort seriously affects the comfort during exercise and hinders limb movement (Ashdown 2011), such as frequent contraction of the human body's joint parts while doing sports (Fan 2009) and displacement of the female breast (Zhou et al. 2011; Zhou et al. 2012). Moreover, the cut and sew technique based on one ready fabric caused up to 40% of waste (Carvalho et al. 2015).

Apparel manufacturing is labour intensive (Scott 2006), especially in the sewing process, requiring a lot of workforce. Although some 3D knitting techniques are used in the production of sportswear and underwear, they still need to be cut and sewn for obtaining shapes which cause 10%–15% material waste (Troynikov and Watson 2015). These sustainable issues and ergonomic fitting issues requirements disclose the needs of seamless forms with tailored tactility and functional areas.

This research aims to establish seamless pattern geometry from 3D anthropometry captured by a contact-less body scanning system for the purpose of 3D weaving of seamless sports bra employing cross-platform digital technology. The latest 3D weaving technology permits the interlocking of the yarns in the X (longitudinal-warp), Y (cross-weft) and Z (vertical-through-thickness) directions (Shi et al. 2021). These three interlocking directional yarns are then tailored within the amalgamated weave architectures completed in one weaving cycle for a seamless form. Bio-tech-composite production fuses both human and digital manufacturing technology mechanics-motions underpinned by anthropometry and ergonomic real-time data. The whole 3D-to-2D-to-3D process was identified as two parts—3D complex curved body surface geometries converted into a flattened 2D graphical pattern (3D-to-2D) using a cross-platform CAD system, and 2D geometries with appropriate weave architectures transmitted back to 3D seamless woven spherical composites (2D-to-3D)—while taking-off the loom. The research and development into seamless woven bra forms utilise such data to produce variable spherical forms that mould and support a given set of 3D geometries. These 3D geometries are translated into a 2D form within a graphical template. Through the conversion process, 3D-to-2D, design parameters need to be included ensuring a concise number of warps and wefts (columns and rows within the graph template). Upon completion of weaving this 2D design in the 2D fabric plane and removed from the constraints of the weaving machine, it is then converted back into a 3D shape by pulling-pushing-unfolding the woven fabric. Associated to multilayer (warp), multilevel (weft) woven fabric enables specific drape, support and movement. All manufacturing processes are completed in one weaving cycle, which reduces the post-manpower involved. A hybrid of fibre-yarn types within the compositing of the 3D woven seamless fitting bra enables the optimisation of such undergarments for physical-active sportswear use. This process bridges the clothing field with the latest technical textile manufacturer technology. Positioning the work within the sportswear–sports bra market enables a framework for the range of generic design geometries.

5.2 MATERIALS AND METHODS

The 3D-to-2D-to-3D process in this research is shown in Figure 5.1. The 3D geometries based on real-time ergonomic data-driven are converted into 2D graphical schematics with segmentations and artificial boundary lines. The various multi-application digital systems employed the process ensure to obtain the accurate 3D and 2D parametric geometry database. The ergonomic data-driven consideration with the help of a cross-platform system would enhance and develop technical manufacturing technology.

The 3D surface geometry of the SS female sample (Size Stream, LLC) was captured by Size Stream SS20 Booth Scanner (Size Stream, LLC, USA) and processed with a reverse engineering system—Germanic Warp (3D Systems, Inc., GW) initially in .obj data format. It was further processed in GW to reconstruct and repair the 3D geometry of the bust shape. The 2D pattern blocks were flattened in CLO 3D (CLO Virtual Fashion Inc.), based on the repaired and reconstructed 3D digital mannequin that was imported into CLO 3D as the avatar. This converted 2D graphical schematic was additionally inputted in BOK Garment CAD System (Shenzhen BOK Investment Co., Limited., BOK, V13) with marked reference points. Finally, EAT 3D WEAVE COMPOSITE (EAT GmbH, "The DesignScope Company", EAT 3D) was applied to draw the 2D pattern filled with multilayer and multilevel weaving architectures fed into the Mageba multishuttle weaving machine and Staubli UNIVAL 100 jacquard harness (MS-100).

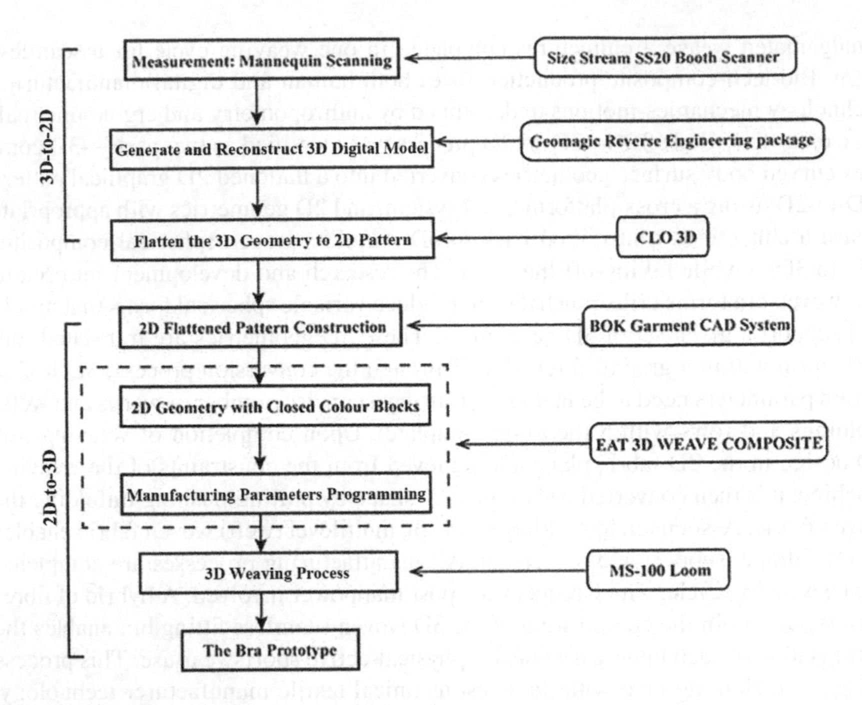

FIGURE 5.1 The 3D-to-2D-to-3D research process workflow.

5.2.1 The 3D-to-2D Geometry Process

The high-performance booth scanner SS20 (Size Stream, USA) with automatic and 20 safe infrared depth sensors was exploited in this research. The scanned data was obtained as a .obj format file. This 3D digital data (.obj file) was imported into the Geomagic Wrap (3D Systems, Inc.) to generate the digital model.

Geomagic Wrap is a 3D digital model analysis software with advanced surfacing tools (3D Systems, Inc.). The polygon meshes firstly examined the scanning model in the reverse-engineering process. The "Mesh Doctor" function (see Figure 5.2) initially corrected the faulty mesh data automatically, and the utilisation of "Polygons & Curvature" precisely repaired the 3D model. The "Trim with Plane" function was used to select and cut off useless model parts.

After pre-processing, the scanning model was imported into the CLO 3D as the avatar. As the bra is a symmetrical garment, the whole bra was consequently obtained through the symmetrical copy of the left-side pattern. The "3D Pen (Avatar)" drew the 3D pattern on the avatar (digital scanning model) surface, then "Edit 3D Pen (Avatar)" was used to adjust the line details operating the function—"Add or Deleted (Curved) Points". Before using the "Flatten" tool, the middle centre line and side line were labelled by "Flattening as the Straight Line" tool. All 3D geometry pattern were selected and highlighted in the 3D window, and the "Flatten" tool was used to flatten the 3D solid geometry to the 2D sheet pattern. The following "Edit Pattern (Z)", "Edit Curvature (C)" and "Edit Sewing" devices were supplementarily used to

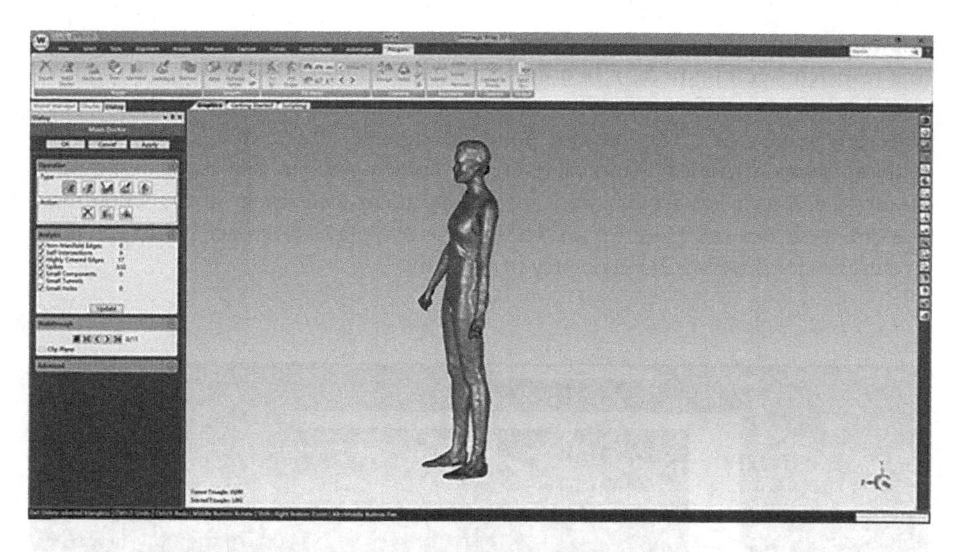

FIGURE 5.2 The screenshot of GW Mesh Doctor function.

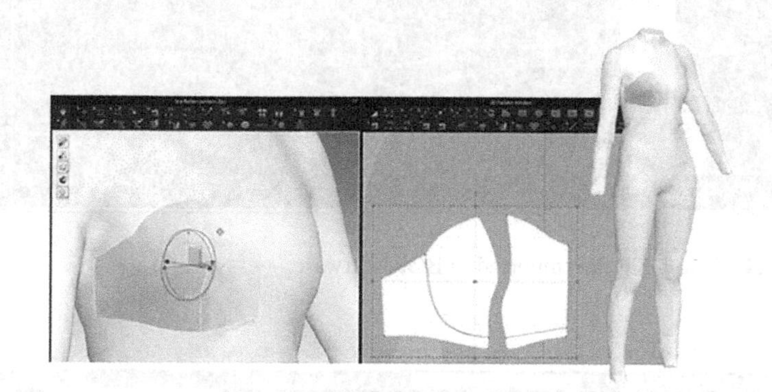

FIGURE 5.3 The 3D and 2D visual pattern in CLO 3D.

amend the 2D sheet pattern. The result in CLO 3D is shown in Figure 5.3, and the flattened 2D pattern was saved as the DXF file for the further 2D-to-3D process.

5.2.2 The 2D-to-3D Process

5.2.2.1 The 2D Weaving Segment Construction

BOK CAD System is a highly integrated parametric garment pattern making software. Design Mode, Pattern Mode, and Clothing Grading and Marking Mode are three crucial functions. It is also fully compatible with variations of DXF files without distortion, which enables transformations among different software.

The obtained flattened 2D pattern in CLO 3D was saved as the DXF file, and consequently imported into the BOK system, as shown in Figure 5.4. Measuring tools, line plotting and mark points in Intelligence Design Mode were mainly applied in this research (see Figure 5.5). In order to facilitate quantitative drawing in the EAT weaving software, reference points were marked and measured in each 2D flattened pattern boundary line. The marking point function was used to find cross points and aliquots segment points. Amount reference point positions (the distance between points and X-axis or Y-axis) were located by measuring tools in the Coordinate System, as shown in Figure 5.6 (a). These quantity data of reference points provided a reference for drawing 2D geometry in EAT 3D.

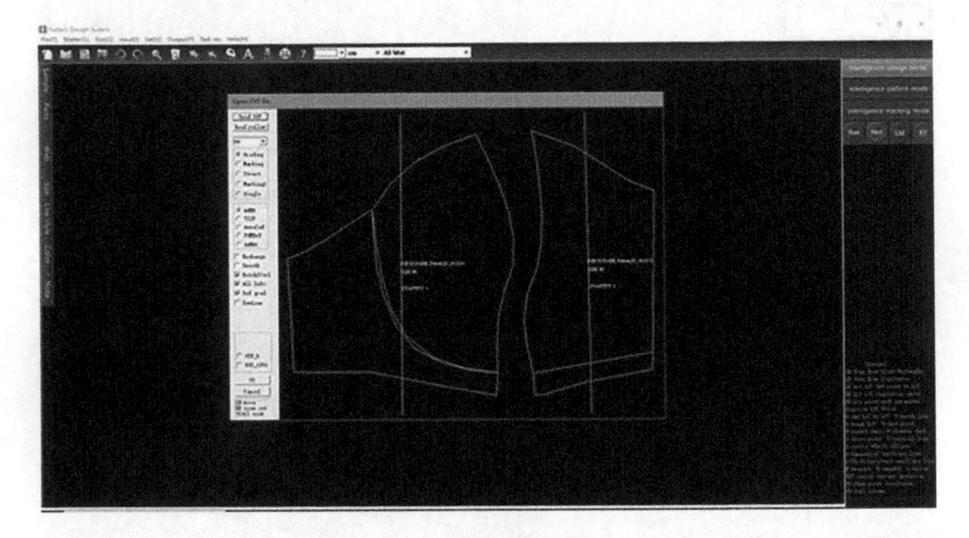

FIGURE 5.4 The DXF file imported in BOK software.

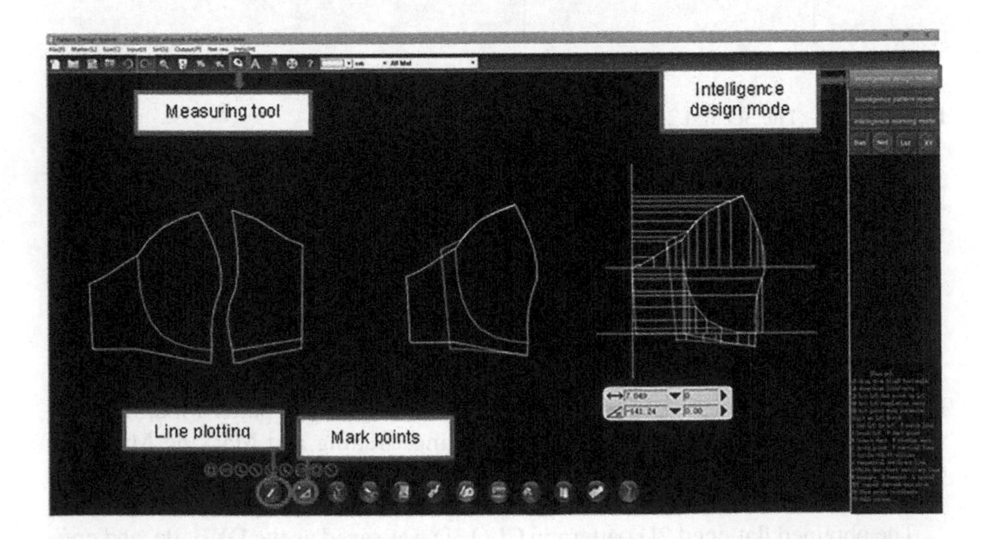

FIGURE 5.5 The screenshot of operation in BOK system.

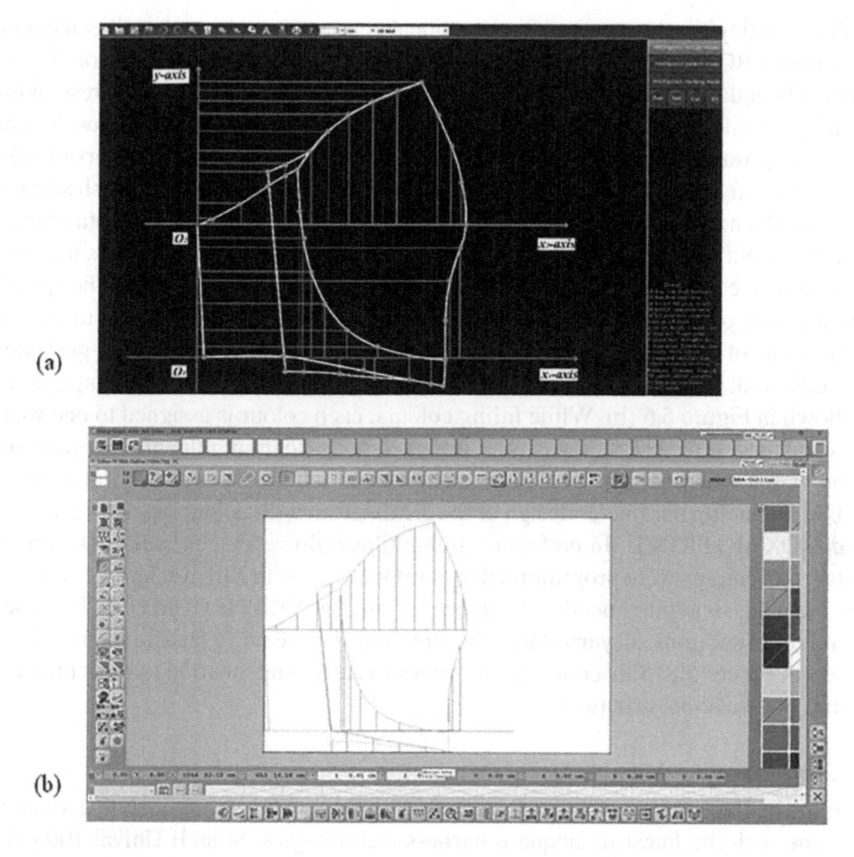

FIGURE 5.6 (a) Reference points making and measuring in BOK; (b) The reference points and 2D pattern boundary lines drawn in EAT 3D.

EAT 3D is designed for technical weaving. It translated the 2D geometry with manufacturing information and programmed production parameters, for instance, shuttle(s) movement, take-up/let-off speed, into JC5 files for feeding into weaving loom. In EAT 3D, these functions were principally exploited to arrange to manufacture parameters.

- EDITOR IV is an area to draw design pictures based on real-time data according to CAD images in BOK.
- BOX MOTION II is an area to design zones and zone programming; this function is used to design producing parameters communicating with MEGABA.
- 3D WEAVE COMPOSITE is used to create weaving structures and composite with the help of cross-section design in weft and warp visually.
- ASSIGN WEAVES TO COLOURS: weaves will be assigned to colours of design using existing or generating new weaves.

After getting the drawings of conversational geometry results with artificial boundary lines in BOK, EAT 3D was used to draw closed colour blocks based on the calculated boundary line, subsequently, colour blocks were filled with corresponding multilayer, multilevel weaving architectures. All 2D geometric conversion results were drawn by primary functions in EDITOR IV. Figure 5.6 (a) shows the marked reference points in BOK software. Auxiliary coordinate systems were established based on Point O_1 and Point O_2 as the origin points, respectively. The location data of every reference point was measured using BOK measuring tools. Based on BOK's measurement results, each reference point is located by the line tool in EAT 3D. The specific dynamic size of the line is shown at the bottom of the design area. Curve tools, with applications of curve fill tools for larger areas and the brush tool for each grid, were utilised to link located reference points to make closed segments before filling colours, as shown in Figure 5.6 (b). While filling colours, each colour is assigned to one weaving structure, so the colour areas were designed consistent with the arrangement area of the woven structure. The final 2D pattern with colour blocks is shown in Figure 5.7.

The producing parameter design of the 2D diagram with colour blocks was carried on in BOX MOTION II. In preference to multilayer dome shapes, roller take-up and shuttle movements were programmed to communicate with MS-100 loom. Moreover, new weaving structures need to be generated in 3D WEAVE COMPOSITE. There are related functions of yarn editor for creating the weaving structure visually, as shown in Figure 5.8. Subsequently, 3D Viewer is a decent function to detect the correctness of weaving structure.

5.2.2.2 Weaving Manufacture Process

Combining the latest in advanced weaving machines, Mageba multishuttle weaving machine with the latest in jacquard harness technologies, Staubli Unival 100 (MS-100) (see Figure 5.9), enabled the successful stages of geometric flattening of a given

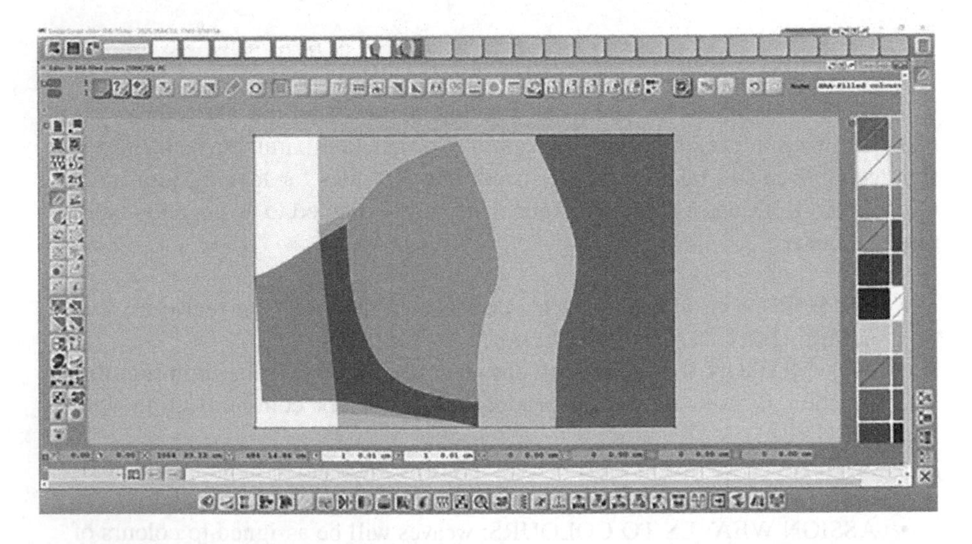

FIGURE 5.7 The screenshot of 2D flatten pattern drawn in the EAT 3D software.

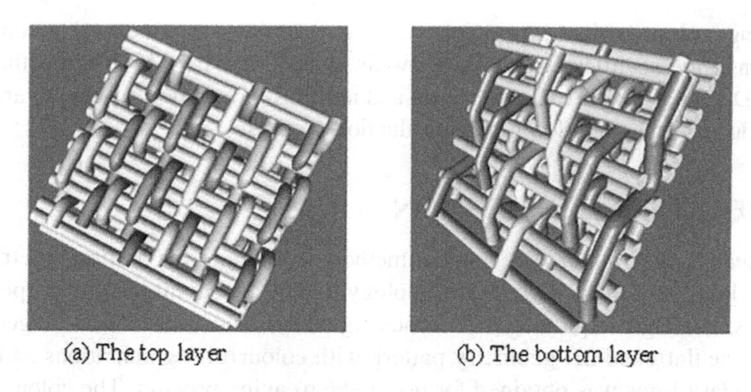

(a) The top layer (b) The bottom layer

FIGURE 5.8 The 3D version of double layer twills.

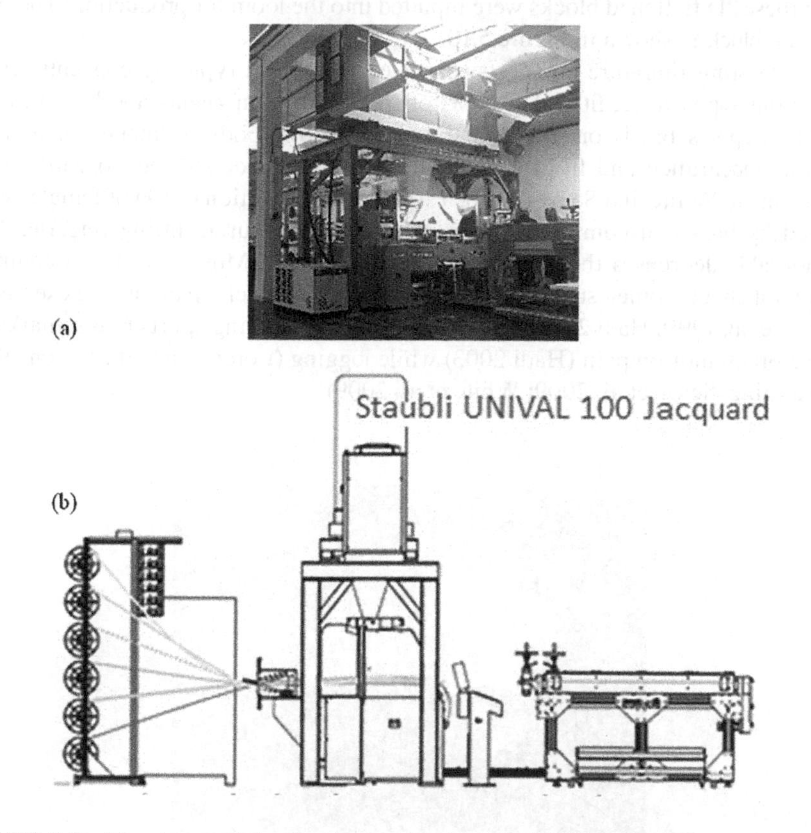

FIGURE 5.9 Mageba multishuttle weaving machine and Staubli UNIVAL 100 jacquard harness.

form for design for production to product/prototype: 3D-to-2D-to-3D woven spherical forms (Shi et al. 2021). The inclusion of advancing weaving machine and harness mechanics and motions of the MS-100 in the production parameters is included within the design stages relative to the weaving cycle. This permits developments in advancing on loom origami principles to more complex components that comprises in this instance seamless woven sportswear/sports bra. This MS-100 together with 3D-to-2D-to-3D software design stages and multilayer, multilevel weaving architectures reduces further off-loom post-production processes.

5.3 RESULTS AND DISCUSSION

This research innovatively presented a method of bridging the anthropometric data with the latest technical weaving technology to develop seamless woven sports bra cup areas. Through employing the various digital systems in the 3D-to-2D geometry process, the flattened 2D geometry pattern with colourful segmentations and artificial boundary lines was obtained for use in the weaving process. The colour blocks are filled with the weaving information containing the weaving architectures and lines; these 2D flattened blocks were inputted into the loom for production. The final bra front block is shown in Figure 5.10.

The existing literature on sportswear is extensive and typically concentrates on improving appearance, fit and comfort of sports bra components and breast movements. A sports bra is one kind of the closest fitting body garments for females and the appearance and fit of that are demanding (Hardaker and Fozzard 1997). According to White and Scurr (2012), a significant proportion (69%) of females were swayed by breast discomfort because of wearing inaccurate fitting lingerie. This considerably decreases their health and quality of life. Moreover, it is a common complaint about women suffering from cyclic and non-cyclic pain during exercising (Mason et al. 1999; Hadi 2003; Zhou et al. 2011). Well-fitting sports bras remarkably relieve breast motion pain (Hadi 2003) while jogging (Lorentzen and Lawson 1987) and running (Scurr et al. 2009; White et al. 2009).

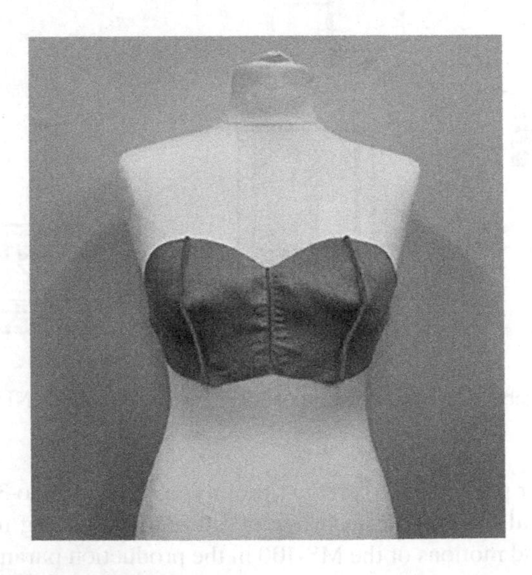

FIGURE 5.10 The final front block and the woven bra sample.

3D digital tools and 3D scanners are predominantly applied in the bra design process to improve the fitting issues (Coltman et al. 2018). The two auxiliary criteria, BMI and age, were suggested to assist consumers in choosing well-fitting bras after examining the 3D scanning variables (Shi et al. 2020). With the aid of the 3D scanner, errors in scanning the breast volume of women with large breasts in a standing posture were found (Coltman et al. 2017); this further improves the bra design for females with bigger cups. Besides the 3D scanning, the 4D scanning technology was used to measure the vertical displacement and shape changes of breasts during running (Pei et al. 2021), laying a foundation for breast kinematics and sports bras with better functional basis. In agreement with the earlier design process, the woven bra pattern was gathered on the basis of the original 3D scanning model directly instead of the existing bra, which contributes to reducing the pattern design errors and producing a better fitting bust shape.

In addition to 3D digital platforms, the cross-use of 2D/3D CAD/CAM systems, aimed in various clothing and textile software, smoothly links the garment pattern design procedure with advanced manufacturing technology to develop well-fit apparel wear. Prior studies have also noted the significance of the mixed 3D and 2D digital platform application in optimising traditional garment design routines. Dāboliņa et al. (2018) present a method of automatically applying anthropometric data in CAD/CAM systems to enhance automated garment design and daily production processes. Shin et al. (2010) propose converting 3D geometry to 2D geometry based on linear least squares optimisation in order to shorten the calculation time of physical methods and improve the efficiency of patternmakers and designers. Such studies still focussed on the traditional manufacturing technique; however, few studies explored ways to introduce advanced manufacture technology applied in the apparel industry.

The drawback of the cut and sew technique is obvious, high material and cost waste, labour intensive (Nayak and Padhye 2015; Troynikov and Watson 2015) and weaker protection and performance (Chen and Yang 2010a). There is literature investigating zero-waste design as well. The use of as little cutting as possible to make finished garments while taking into account the aesthetics and function of the garments are demonstrated (Carrico and Kim 2014); the works of McQuillan (2020) practised and emphasised the vital role of 3D software in the zero-waste fashion design process. Compared with 2D software, 3D software has more robust visualisation and simulation, finding problems promptly and saving time. Although efforts have made some progress in maximum waste reduction, the drawbacks of traditional production methods are still tricky. New production technology—3D weaving technology—can solve some shortcomings of conventional production technology and also initially applied protective wear—female body armour. The previous studies explore ways to apply the 3D warp interlock fabric and angle-interlock fabrics in the seamless female body armour (Chen and Yang 2010a, 2010b; Abtew et al. 2018, 2020). These technical fabric applications with significant ballistic and fitting improvement show the possibility to introduce 3D weaving into the clothing industry. Consistent with this, this research innovatively explores the 3D technical weaving in the underwear process to improve the fitting issues with a more sustainable manufacturing process.

5.4 CONCLUSION

The focus of this research was to develop a composite hybrid form containing fibre-yarn blends and multilayer (warp) and multilevel (weft) to interlock and be shaped in one cycle for well-fitting apparel wear, protection and performance. The final result also confirmed the feasibility of this method. In this research, the cross-platform software technology—3D reverse engineering system, 2D CAD clothing system and technical textile CAD/CAM—were exploited to bridge 3D anthropometry with the latest textile technology to develop seamless woven sports bras. This highlights the importance of the digital platform in cross-professional research. The comprehensive cross-platform digital tools application provides more guarantee for cross-professional/cross-field research. This research methodology would be beneficial to produce the 3D spherical composites, not limited to the sports bra. This 3D-to-2D-to-3D bra procedure is also likely to be explored in other spherical shape clothing parts—located in areas such as the hips, elbows, and knees—and other end textile production applications.

REFERENCES

Abtew, Mulat Alubel, Francois Boussu, Pascal Bruniaux, Carmen Loghin, and Irina Cristian. 2020. "Enhancing the ballistic performances of 3D warp interlock fabric through internal structure as new material for seamless female soft body armor development." *Applied Sciences (Switzerland)* 10 (14): 4873. https://doi.org/10.3390/app10144873.

Abtew, Mulat Alubel, Pascal Bruniaux, François Boussu, Carmen Loghin, Irina Cristian, and Yan Chen. 2018. "Development of comfortable and well-fitted bra pattern for customized female soft body armor through 3D design process of adaptive bust on virtual mannequin." *Computers in Industry* (100) (March): 7–20. https://doi.org/10.1016/j.compind.2018.04.004.

Ashdown, Susan P. 2011. "Improving body movement comfort in apparel." In *Improving Comfort in Clothing*, edited by G. Song, 278–302. Cambridge: Woodhead Publishing Limited in association with The Textile Institute.

Bartels, Volkmar T. 2006. "Physiological comfort of sportswear." In *Textiles in Sport*, edited by Roshan Shishoo, 43–44. Cambridge: Woodhead Publishing Limited in association with The Textile Institute. https://doi.org/10.1201/9781439823804.ch9.

Carrico, Melanie, and Victoria Kim. 2014. "Expanding zero-waste design practices: A discussion paper." *International Journal of Fashion Design, Technology and Education* 7 (1): 58–64. https://doi.org/10.1080/17543266.2013.837967.

Carvalho, M., H. Carvalho, L.F. Silva, and F. Ferreira. 2015. "Sewing-room problems and solutions." In *Garment Manufacturing Technology*, edited by Rajkishore Nayak and Rajiv Padhye, 317–336. Amsterdam: Elsevier.

Chen, Xiaogang, and Dan Yang. 2010a. "Use of 3D angle-interlock woven fabric for seamless female body armor: Part 1: Ballistic evaluation." *Textile Research Journal* 80 (15): 1581–1588. https://doi.org/10.1177/0040517510363187.

Chen, Xiaogang, and Dan Yang. 2010b. "Use of three-dimensional angle-interlock woven fabric for seamless female body armor: Part II: Mathematical modeling." *Textile Research Journal* 80 (15): 1589–1601. https://doi.org/10.1177/0040517510363188.

Colovic, Gordana. 2015. "Sewing, stitches and seams." In *Garment Manufacturing Technology*, edited by Rajkishore Nayak and Rajiv Padhye, 247–273. Amsterdam: Elsevier. https://doi.org/10.1016/B978-1-78242-232-7.00010-2.

Coltman, Celeste E., Deirdre E. McGhee, and Julie R. Steele. 2017. "Three-dimensional scanning in women with large, ptotic breasts: Implications for bra cup sizing and design." *Ergonomics* 60 (3): 439–445. https://doi.org/10.1080/00140139.2016.1176258.

Coltman, Celeste E., Julie R. Steele, and Deirdre E. McGhee. 2018. "Which bra components contribute to incorrect bra fit in women across a range of breast sizes?" *Clothing and Textiles Research Journal* 36 (2): 78–90. https://doi.org/10.1177/0887302X17743814.

Daanen, Hein A.M., and Frank B. Ter Haar. 2013. "3D whole body scanners revisited." *Displays* 34 (4): 270–275. https://doi.org/10.1016/j.displa.2013.08.011.

Daanen, Hein A.M., and Jeroen Van De Water. 1998. "Whole body scanners." *Displays* 19 (3): 111–120. https://doi.org/10.1016/s0141-9382(98)00034-1.

Dāboliņa, Inga, Ausma Viļumsone, Jānis Dāboliņš, Eugenija Strazdiene, and Eva Lapkovska. 2018. "Usability of 3D anthropometrical data in CAD/CAM patterns." *International Journal of Fashion Design, Technology and Education* 11 (1): 41–52. https://doi.org/10.1080/17543266.2017.1298848.

Fan, Jintu. 2009. "Physiological comfort of fabrics and garments." In *Engineering Apparel Fabrics and Garments*, edited by Jintu Fan and Lawrance Hunter, 201–250. Cambridge: Woodhead Publishing Limited in association with The Textile Institute.

Hadi, Maha S.A. Abdel. 2003. "Sports brassiere: Is it a solution for mastalgia?" *The Breast Journal* 6 (6): 407–409. https://doi.org/10.1046/j.1524-4741.2000.20018.x.

Hardaker, Carolyn H.M., and Gary Fozzard. 1997. "The bra design process: A study of professional practice." *International Journal of Clothing Science and Technology* 9 (4–5): 311–325. https://doi.org/10.1108/09556229710175795.

Iersel, Miranda van, Henny Veerman, and Wannes van der Mark. 2009. "Modelling a crime scene in 3D and adding thermal information." *Electro-Optical and Infrared Systems: Technology and Applications* 6 (7481): 74810M. https://doi.org/10.1117/12.829990.

Kennedy, Kate. 2015. "Pattern construction." In *Garment Manufacturing Technology*, edited by Rajkishore Nayak and Rajiv Padhye, 205–220. Amsterdam: Elsevier. https://doi.org/10.1016/B978-1-78242-232-7.00008-4.

Kunz, Grace, and Ruth Glock. 2004. *Apparel Manufacturing: Sewn Product Analysis (Fashion Series)*, 4th ed. Upper Saddle River, NJ: Pearson/Prentice Hall.

Lorentzen, Deana, and LaJean Lawson. 1987. "Selected sports bras: A biomechanical analysis of breast motion while jogging." *Physician Sports Medicien* (15): 128–139. https://doi.org/10.1080/00913847.1987.11709355

Mason, Bruce R., Kelly-ann Page, and Keiran Fallon. 1999. "An analysis of movement and discomfort of the female breast during exercise and the effects of breast support in three cases the three subjects who participated in the study were active young women who." *Journal of Science and Medicine in Sport* 2 (2): 134–144.

McQuillan, Holly. 2020. "Digital 3D design as a tool for augmenting zero-waste fashion design practice." *International Journal of Fashion Design, Technology and Education* 13 (1): 89–100. https://doi.org/10.1080/17543266.2020.1737248.

Mochimaru, Masaaki, and Makiko Kouchi. 2011. "4D measurement and analysis of plantar deformation during walking and running." *Footwear Science* 3 (1): S109–S112. https://doi.org/10.1080/19424280.2011.575878.

Nayak, Rajkishore, and Rajiv Padhye. 2015. "Introduction: The apparel industry: The apparel industry." In *Garment Manufacturing Technology*, edited by Rajkishore Nayak and Rajiv Padhye, 1–17. Amsterdam: Elsevier. https://doi.org/10.1016/B978-1-78242-232-7.00001-1.

Pei, Jie, Linsey Griffin, Susan P. Ashdown, and Jintu Fan. 2021. "The detection of the upper boundary of breasts using 4D scanning technology." *International Journal of Fashion*

Design, Technology and Education 14 (1): 1–11. https://doi.org/10.1080/17543266.2020.1829097.

Robinette, Kathleen M., and Hein A.M. Daanen. 2006. "Precision of the CAESAR scan-extracted measurements." *Applied Ergonomics* 37 (3): 259–265. https://doi.org/10.1016/j.apergo.2005.07.009.

Sayem, Abu Sadat Muhammad. 2017. "Objective analysis of the drape behaviour of virtual shirt, Part 2: Technical parameters and findings." *International Journal of Fashion Design, Technology and Education* 10 (2): 180–189. https://doi.org/10.1080/17543266.2016.1223810.

Sayem, Abu Sadat Muhammad, Richard Kennon, and Nick Clarke. 2010. "3D CAD systems for the clothing industry." *International Journal of Fashion Design, Technology and Education* 3 (2): 45–53. https://doi.org/10.1080/17543261003689888.

Scott, Allen J. 2006. "The changing global geography of low-technology, labor-intensive industry: Clothing, footwear, and furniture." *World Development* 34 (9): 1517–1536. https://doi.org/10.1016/j.worlddev.2006.01.003.

Scurr, Joanna, Jennifer White, and Wendy Hedger. 2009. "Breast displacement in three dimensions during the walking and running gait cycles." *Journal of Applied Biomechanics* 25 (4): 322–329. www.ncbi.nlm.nih.gov/pubmed/20095453.

Shi, Yuyuan, Hong Shen, Lindsey Waterton Taylor, and Vien Cheung. 2020. "The impact of age and body mass index on a bra sizing system formed by anthropometric measurements of sichuan Chinese females." *Ergonomics* 63 (11): 1434–1441. https://doi.org/10.1080/00140139.2020.1795276.

Shi, Yuyuan, Lindsey Waterton Taylor, Vien Cheung, and Abu Sadat Muhammad Sayem. 2021. "Biomimetic approach for the production of 3D woven spherical composite applied in apparel protection and performance." *Applied Composite Materials*, no. 0123456789. https://doi.org/10.1007/s10443-021-09936-5.

Shin, Kristina, Sun Pui Ng, and Ma Liang. 2010. "A geometrically based flattening method for three-dimensional to two-dimensional bra pattern conversion." *International Journal of Fashion Design, Technology and Education* 3 (1): 3–14. https://doi.org/10.1080/17543260903460200.

Troynikov, Olga, and Chistopher Watson. 2015. "Knitting technology for seamless sportswear." In *Textiles for Sportswear*, edited by Roshan Shishoo, 95–117. Cambridge: Woodhead Publishing. https://doi.org/10.1016/B978-1-78242-229-7.00005-9.

White, Jack Lee, and Joanna Scurr. 2012. "Evaluation of professional bra fitting criteria for bra selection and fitting in the UK." *Ergonomics* 55 (6): 704–711. https://doi.org/10.1080/00140139.2011.647096.

White, Jack Lee, Joanna Scurr, and Neville Horton Smith. 2009. "The effect of breast support on kinetics during overground running performance." *Ergonomics* 52 (4): 492–498. https://doi.org/10.1080/00140130802707907.

Zhou, Jie, Winnie Yu, and Sun Pui Ng. 2011. "Methods of studying breast motion in sports bras: A review." *Textile Research Journal* 81 (12): 1234–1248. https://doi.org/10.1177/0040517511399959.

Zhou, Jie, Winnie Yu, and Sun Pui Ng. 2012. "Studies of three-dimensional trajectories of breast movement for better bra design." *Textile Research Journal* 82 (3): 242–254. https://doi.org/10.1177/0040517511435004.

Part C

Digital Human and Metaverse

6 Processing Data from High Speed 4D Body-Scanning System for Application in Clothing Development

Yordan Kyosev, Vanda Tomanova and Tatjana Spahiu

CONTENTS

6.1 INTRODUCTION

Clothing products are used in everyday life. They support thermal regulation of the human body and protect from heat or cold (Gilligan 2016). They are categorised for

DOI: 10.1201/9781003264958-9

different uses or purposes, such as for use as daily wear and applications in sport, medicine, aeronautics, etc., and require different properties. Despite the purpose of clothing manufacture, construction plays an important role in creating products that offer consumers the right fit and appearance. Obviously, they are strongly related to materials used for fabric production. Expectations from garments that are used for a specific purpose are high. For example, athletes require different features of clothing, such as minimal air resistance, low levels of sweating, little to no disturbance in motion, and ideally even muscular support.

Compression products are among other products that require accurate body measurements in order to provide the right pressure on the surface of the individual body part. Workwear used in specific professions requires not only to provide effective protection to wearers, but also to ensure good freedom of movements. Design of these types of clothing is mainly based on body measurements taken in a static state of the human body without taking the changes of the body shape during its motion into account. The automatic measurement of the characteristic body circumferences and lengths using 3D scanning for predefined static poses (like A Pose) are nowadays well established in the scanning systems. The change of body geometry during motion is a new area of research. The difficulty faced during body measurement in motion is the detection of the specific geometrical body positions. However, this should be taken into consideration and used by pattern makers in order to produce garments with the right fit and comfort (Bragança et al. 2016), especially for tight-fitting garments (Tama and Öndoğan 2020).

This chapter demonstrates one possibility for such measurements by using the delivered homologous meshes from Move4D scanning software. These changes in body geometry during motion should be taken and used by pattern makers in order to produce garments with the right fit and comfort.

6.2 OVERVIEW OF THE LATEST DEVELOPMENTS IN 4D BODY-SCANNING TECHNOLOGY

Product development of garment and footwear requires accurate human body measurements. The classical manual process of taking anthropometric data takes a long time to get measurements per person and does not have very good reproducibility. The collection of statistical data required involves a high number of people and is a costly process. In the last decades, digital technologies such as 3D scanning have shown a number of advantages in recording accurate body measurements in a quick and easy way (Rumbo-Rodríguez et al. 2021) as it is more reliable (Parker et al. 2017), reproducible (Medina-Inojosa et al. 2016), and has a lower variance compared to manual measurement methods (Lindell et al. 2021). The 3D body scanning has added value to the clothing industry due to a number of advantages offered by this technology. Different producers implement it in order to improve the buying decision of consumers finding the right fit.

A digital copy of the human body contains all the information about surface geometry. There is a high number of scanning systems available on the market starting from stationary, handheld, and mobile scanners (Barto et al. 2021). Dependent on the type of scanning system used for capturing the 3D surface of a human body, the data form a high number of point clouds. These point data are converted to a mesh, which is further elaborated for smoothing, decimation, hole filling, etc. (Rajulu and Corner 2013). The 3D data of the human body can be exported and used to extract anthropometric data, virtual modelling, and simulation or even predict garment fit, especially for online clothes shopping. This growing market is expected to increase by 9.1% per year until 2025 (Fashion eCommerce report 2021 2021). Due to the returns which are mainly caused by the wrong fit, the costs associated are higher for retailers. Providing information about, for example, the virtual fit may reduce the number of returns (Stöcker et al. 2021). Among various applications of 3D body scanning technology, the virtual fitting room shows the highest rate of growth and is expected to rise from USD 2.9 billion in 2019 to USD 7.6 billion by 2024 (Markets and Markets 2022). Mapping the precise surface of the human body and generating accurate anthropometric data from a 3D scanner leads to an improvement in product development. This can be used for different applications. Especially for military clothing and equipment, the issue of clothing fit is of great importance. In the case of sportswear, which is mainly focused on users' performance and comfort, the customised outfit can improve their performance (Possi 2018). 3D body scanning has proven as well to be a fast and accurate method in taking anthropometric data to develop clothing patterns (Sohn et al. 2020), especially for customized sportswear (De Raeve et al. 2018; Giachetti et al. 2015; Wu and Kuzmichev 2020). 3D scanners allow precise scanning of static positions; however, they lack the ability to capture body movement. As all wearable products are subject to body movement or human motion, latest technologies have introduced 3D scanning in motion or 4D scanning. The "Textile/Clothing and Technology Corporation", well-known in the fashion industry as 3D scanning technology, has recently developed a 4D mode which enables 3D scanning and movement visualization of the human body. It can be used mainly not only in the fashion industry but in medicine and fitness, too (Textile/Clothing and Technology Corporation n.d.).

The Max Plank Institute for Intelligent Systems is using a 4D scanner for full body scanning by the company 3DMD and developed algorithms for evaluation of the data with this scanner. The main scope is to study the human body in motion; data collected can be used for graphics, medicine, psychology, and computer vision and recently focused on 4D motions of clothing on the body (Max Planck Institute for Intelligent Systems 2022).

In addition, another 4D body scanner, Alice, was developed by 3DCopySystems. It was introduced by Hohenstein Institute (Bönningheim, Germany) with the main scope to analyse and improve product development in the apparel industry (Hohenstein launches series of 4D scanning projects 2019). In the study, a group of male participants—either athletes or physical workers—were tested in a static and dynamic state to demonstrate the benefit of 4D scanning technology to analyse and evaluate the interaction between clothing and the human body for two movements, including "Arm Flex" and "Leg Flex" (Klpeser et al. 2021). The authors continued their investigation with three works or sport-related movements that were defined and compared to the standard position: biceps curl, leg flex, and squat (Klepser and Morlock 2020). Meanwhile, other researchers were focused on measuring the human breast in motion. For this purpose, 3DMD body 18.t System was implemented to scan twenty-six females for three circumferential measurements which are mainly used for product development. The result shows the difference that should be taken into account (Pei et al. 2021). The work continued even for seven non-circumferential measurements that complete a gait cycle during running. Their findings present significant differences and suggest 4D scanning as an effective way to evaluate changes in breast shape during physical activities (Pie et al. 2021).

Move4D provided by IBV (Spain, Valencia) is an accurate high-speed 3D motion scanner used to capture the human body surface in motion. It can successfully be used in biomechanical analysis in sports and health-related assessments (Parrilla et al. 2019). Some previous attempts by researchers have implemented Move4D already for body scanning and used it to evaluate garments' simulation in motion (Zhang et al. 2021; Lin et al. 2021; Nicolau et al. 2022; Alemany et al. 2022; Kuehn and Kyosev 2021).

6.3 METHODOLOGY OF 4D HUMAN BODY SCANNING

6.3.1 Hardware

In this chapter, 4D scanning, realized with the Move4D system of IBV, is reported. It is the only system whose hardware and software provides consistently (named from the scanner) homologues mesh, where the vertices keep their correspondence to the relevant body position between the frames. The Move4D is a modular scanning system that uses photogrammetry to obtain human body images. The modules

FIGURE 6.1 View of the 4D scanning system at the ITM of TU Dresden.

TABLE 6.1
Technical Characteristics of the 4D Scanning System at the ITM of TU Dresden

Characteristics	Values
Dynamic scanning volume	2 m × 3 m × 3 m
Resolution	1 mm (special resolution)
Capture frequency	Variable from 1 to 179 fps
Scanning time	1 msec/scan
External synchronism	Trigger and synchro input/output
Capture	Synchronised 3D data and texture
Lighting	Inbuilt lighting system
Outputs	Sequence of watertight mesh (OBJ/PLY) with a density of 50,000 points including texture.
	Point cloud (PLY) with a density of about 4 700 000–5 000 000 vertices in the case of one scanned human body.
	Set of 93 automatic body measurements.

are mounted around the scanning space in order to obtain the best visibility of the moving object and are triggered by the processing software. Depending on the scanning width, systems with twelve (2 m × 2 m × 3 m) and sixteen (2 m × 3 m × 3 m) modules are investigated. Each module is equipped with two infrared cameras, one RGB camera, one infrared pattern projector and a personal computer that processes images and synchronises the signals. Figure 6.1 shows the 4D scanning system at the Chair of Development and Assembly of Textile Products, Institute of Textile Machinery and High Performance Material (ITM) of Technical University in Dresden, Germany. Table 6.1 depicts a system description, where main characteristics, their values, and notes are included.

6.3.2 System Calibration

Calibration is an important step to be done before the scanning process to achieve good results in 3D reconstruction. Further, it is very important to establish accurate digital data about the human body. As in every scanning system, there are pre-test methods used to calibrate and test different conditions during the scanning process. Therefore, the Move4D scanning system is equipped with a reference square and a calibration wand. The calibration process itself can be realised in a three-step process.

6.3.2.1 Wand Calibration

Wand calibration is the first step of the system calibration. It consists of sweeping the scanning volume using the wand tool. During this process the modules should detect the LED pulses. This is a real-time process where the total number of frames used, the number of wand frames detected per module, and their locations can be checked.

Calibration wands with active markers are translucid spheres, which are illuminated by green and red LED lights during the calibration process. The wand can be connected to any module. Wand distance and coordinate system elevations are set up. The numbers of minimum and maximum wand frames detected are also selected. Wand calibration shows with which efficiency the LED pulses are detected by the modules. A histogram shows if there are irregularities in the error distribution; if so, a particular module must be recalibrated. When the calibration is successful, the error histogram depicts a general view of the calibration. It is possible to recalibrate a single module. Figure 6.2 presents views from the wand calibration screen.

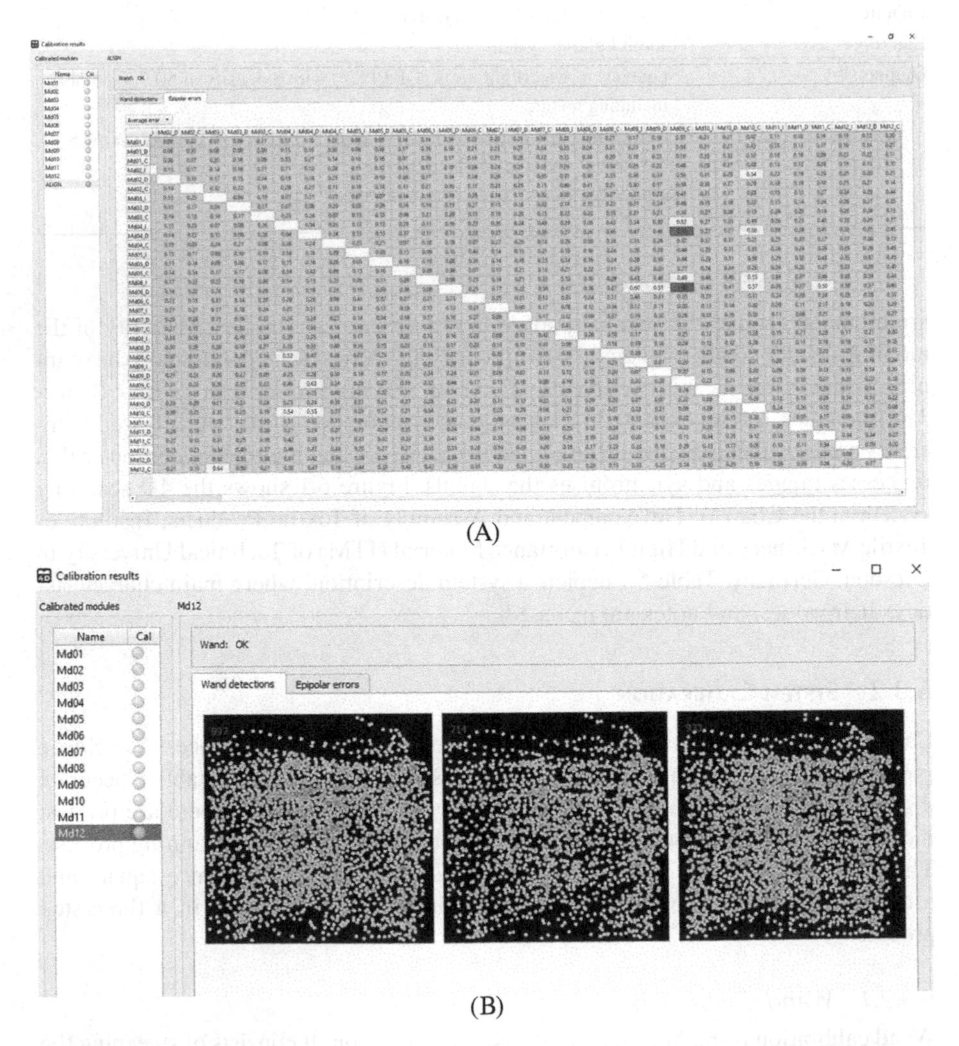

FIGURE 6.2 Wand calibration screen: (A) table with epipolar errors, (B) plot of the detections for cameras of one of the modules, and (C) epipolar errors.

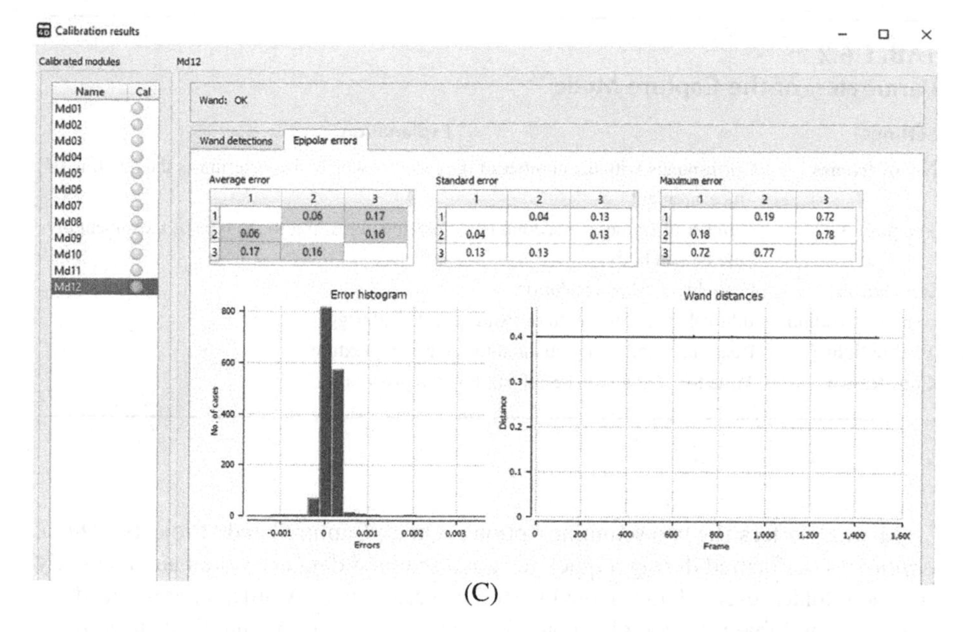

(C)

FIGURE 6.2 (Continued)

6.3.2.2 Calibration of the Coordinate System (CS)

The reference square is placed in the centre of the scanning volume and in parallel to the virtual boundaries. The position of the reference square must be detected afterwards. During this process, the position of the reference square is automatically captured, which is needed for global calibration.

6.3.2.3 Global Calibration

Global calibration is the last step of system calibration. From the result taken from the wand and coordinate system calibration, the automatic algorithm calibrates globally all the modules. Figure 6.2c presents the global epipolar errors and their histograms. On this basis, the calibration of the coordinate system begins, calculating and displaying the global peripolar errors. The calibration process can then be completed.

6.3.2.4 Capture Mode

Here, the users have the possibility to select or create a new pattern of scanning configurations that will be used repeatedly for every subject that will be scanned. To create a new mode, this capture needs to be named and the posture and movement performance needs to be described. Instructions may include information about the subject during the measuring process. Table 6.2 depicts parameters for capture mode that should be filled.

TABLE 6.2
Parameters of the Capture Mode

Settings	Explanation
No. of frames	Corresponds with the number of the capture, which also determines the duration of measurement
Frequency	Number of frames per second (fps) selected according to the type of movement of the human body
Resolution	From high to low resolution
Synchronization	Internal synchronisation or using an external trigger
Visible light	Recommended to be activated to capture the texture
Countdown	Time for preparation before starting the process

Move4D offers the two scanning options "Quick Capture" and "Projects". *Quick capture* is performed during a quick test, and scanned data are stored automatically in a new folder, created and named with the current date. A quick capture needs to create a new subject for which automatic, mandatory, and optional fields need to be filled. After that, the scanning process starts. *Project* is performed during a series of scanning tests by following a protocol that includes different scanning conditions. The project is indeed for registration of different data, and the software automatically names the file; different fields include socio-demographic data and a protocol of scans that can be used as an instruction for performing later scans that are part of this project.

6.4 4D HUMAN BODY SCANNING

A young female volunteer is used as the scanning model. Before starting the process, additional reference points and lines are added on the model to facilitate the process of data elaboration on a specific movement. Points and lines were drawn directly onto the body with a marker. Points are placed on the line intersections and marked in a different colour in order to differ from the lines. The visibility of landmarks is tested at the A pose. Figure 6.3 depicts the subject standing in the A pose during the scanning process in front, back, and side views. The reference points marked on different body parts are also visible in the picture.

The markers placed on the body should be visible on the texture of the scanner. For this purpose, the visibility was tested at the A pose stage. Certain landmarks were not visible, due to the lack of light at certain areas, especially in the underarm region. For the chest circumference, only points were used since the sports bra material did not allow the attachment of more detailed landmarks. Figure 6.4 depicts the texture map generated from the scanning process that includes points and lines that are visible after body scanning.

The problem with non-visible points and lines is observed at the underarm or inner thigh. These parts of the body are not taken during the scanning process. The

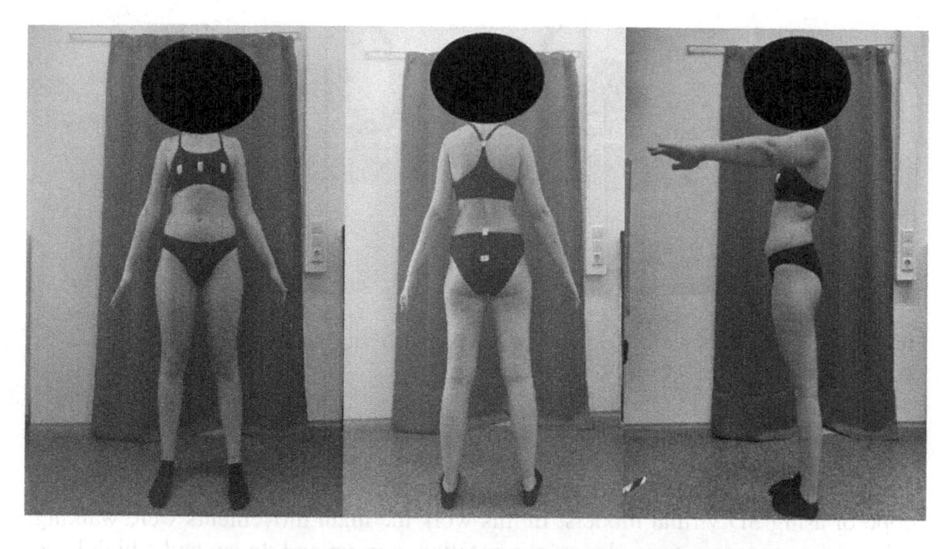

FIGURE 6.3 Front, back, and side views of the subject during scanning process.

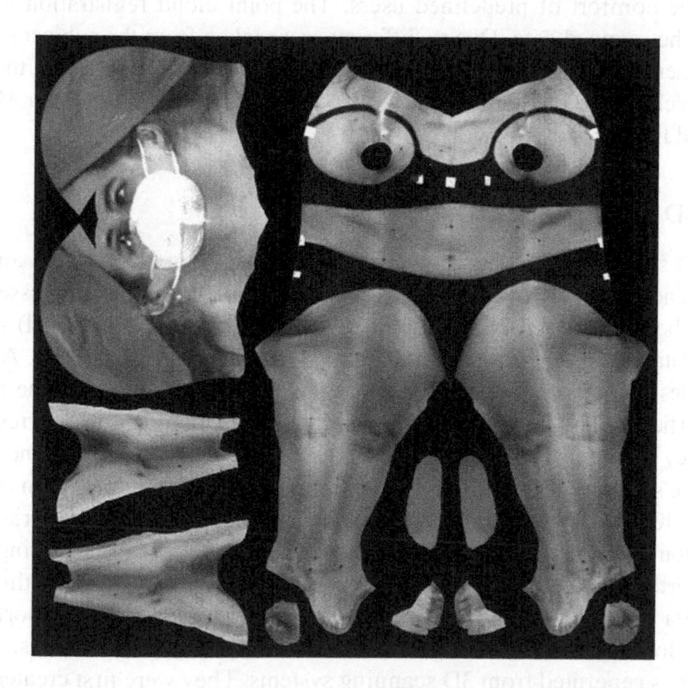

FIGURE 6.4 Texture map of one frame from the body scan.

placement of cameras also creates a "blind spot", where the algorithm does not have sufficient data to provide a satisfactory picture. These spots appear as white on the texture map. This problem was evident in the shoulder area. Figure 6.5 presents some trials undertaken to add landmarks in a post-scanning situation.

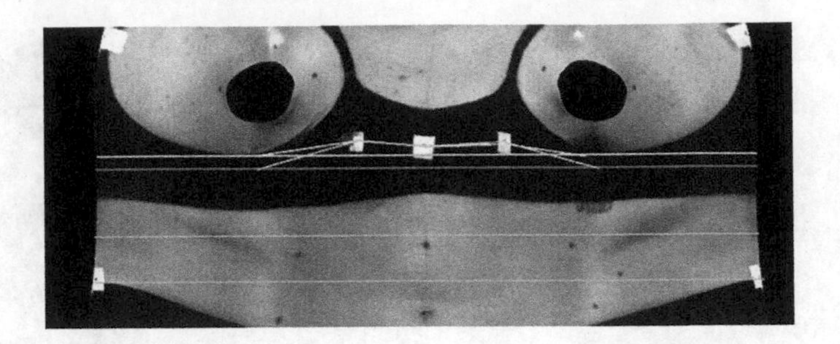

FIGURE 6.5 Trials of adding landmarks in a post-scanning situation.

6.4.1 Selecting Movements

The movement's selection is a procedure that should be included according to the scope of using 3D virtual models. In this work the main movements were walking, sitting down, running, torso down, arm rotation, arm up and down, and a high knee. These movements are selected based on an evaluation for product development to improve the comfort of predefined users. The point cloud registration is used for analysing the scanned data. Due to different scans taken from the subject performing the movements, each one of them represents a fixed point cloud, and the distance between each frame of motions can be calculated. For this purpose, a MATLAB® point cloud library is used.

6.4.2 4D Scan Data Processing

During 4D scanning, the data taken are point clouds by each module and aligned together. The core part of this work uses the homology of the processed meshes of human body. For such scans, a human template (parametric model) is fitted to the scan data and used for the numeration of the vertices in all scans. As a result, all OBJ files, which are processed as homologous mesh, have the same number of points (vertices). The generation of 3D meshes with homologous correspondence (consistency between the frames) along the sequence of scans is done automatically by the software of Move4D. The word *homology* is used mainly in the biology "HOMOLOGUE—the same organ in different animals under every variety of form and function" (Owen 1846; Brigandt 2002). In analogy of the 4D scanning, it means that one vertex corresponds to the same organ or characteristic area of this organ.

The data generated from the Move4D scanning system can be exported in different file formats. The homologue meshes are exported as OBJ files, which are standard files generated from 3D scanning systems. They were first created by Wave front Technologies to save the geometry of 3D objects as lines, polygons, free-form curves, and surfaces (All3DP n.d.). It is a 3D image format that can be exported and imported in various 3D modelling software for editing or additive manufacturing. An OBJ file contains information about texture or colour, which is saved as a separate file with the extension ".mtl" (Material Template Library). This file describes how the 3D programme will apply texture to that object. Moreover, as they are in plain text format, they can be edited in any text editor. As it supports colour and texture information, it is a common file used for 3D printing in different colours. Figure 6.6a shows a list of vertex coordinates that define the geometry of the mesh in x, y, and z coordinates. Meanwhile, Figure 6.6b shows a list of texture coordinates, which are all set between 0 and 1, and f is a list of face elements, defined through vertex, texture, and normal unit vectors (vertex/texture/normal).

(a)

```
 1 # wavefront obj file written by IBV
 2
 3 mtllib A-POSE_0000.mtl
 4
 5 v 0.500787 0.901901 -0.0619354
 6 v 0.500813 0.909635 -0.0619373
 7 v 0.505048 0.903423 -0.065024
 8 v 0.504507 0.911318 -0.0651072
 9 v 0.508806 0.905934 -0.0676007
.0 v 0.499865 0.917414 -0.0626936
.1 v 0.502155 0.918545 -0.0667247
.2 v 0.505281 0.896486 -0.0645973
.3 v 0.501594 0.895381 -0.0605145
.4 v 0.50691 0.891441 -0.0624469
.5 v 0.510379 0.89331 -0.0662185
.6 v 0.49766 0.900802 -0.0573758
.7 v 0.499041 0.894317 -0.055789
.8 v 0.495269 0.900432 -0.052487
.9 v 0.504049 0.890327 -0.0579234
```

(b)

```
100253 vt 0.694336 0.0495605
100254 vt 0.0358887 0.0888672
100255 vt 0.93457 0.0908203
100256 vt 0.699463 0.124268
100257 vt 0.739746 0.0734863
100258 vt 0.415771 0.603516
100259 f 1/1 2/2 3/3
100260 f 3/4 4/5 5/6
100261 f 2/2 6/7 4/8
100262 f 4/5 7/9 133/10
100263 f 1/1 3/3 8/11
100264 f 9/12 8/11 10/13
100265 f 3/4 5/6 143/14
100266 f 8/15 143/14 11/16
```

FIGURE 6.6 Lists of vertex coordinates in (a) first and (b) second part of OBJ file.

6.5 A MATLAB® SCRIPT FOR AUTOMATIC ANALYSIS OF BODY MEASUREMENTS

After the motion scanning, a list of usually about 100 OBJ files with homologous vertices is available for the few seconds of the motion. For their automatic processing, a MATLAB® script was written. It has the task to compute the lengths of pre-defined paths and after that use this length to check their changes during the motions. The distance between predefined points in sequence meshes can be treated as a geodesic distance, which represents the shortest distance between two points (Polthier and Schmies 2006).

At first, the surface is separated into a mesh which is made from a grid of polygons of equal size that approximate the desired surface. Separating a curved surface into a number of flat surfaces can be difficult because the number of them must allow to still follow the curved surface. Most surfaces cannot be flattened without a certain level of approximation that leads to the distorting of given distances. The level of distortion is dependent on the number of polygons into which the object is separated. This separation divides a curve into a number of straight lines. The accuracy of the polyhedral solid is dependent on the original surface. Scanned objects are generally less accurately represented by the polyhedral surfaces than designed objects (Sarkar and Deyasi 2019).

The distance between two points, in this case two vertices characterized by Cartesian coordinates, can be calculated by the distance formula. This formula states that the distance between two points is the square root of the sum of the squares of the differences between corresponding coordinates (Quadricsurface1.docx 2009)

$$\text{Points: } P_1 = (x_1, y_1, z_1),\ P_2 = (x_2, y_2, z_2)$$

Distance between points:

$$d(P_1, P_2) = \sqrt{(x_2 - x_1)^2 - (y_2 - y_1)^2 - (z_2 - z_1)^2}$$

The complete curve length then is the sum of the distances between the:

$$L(P_{1,2,..N}) = \sum_{i=2}^{N} d(P_i, P_{i-1})$$

6.5.1 Preparation of INI File

The analysis is performed in several steps. The first step is the preparation for analysis. During this step the information about the curves, which has to be analysed, is stored in a suitable structure. During the second step, this information is used after reading each OBJ file with the mesh, and only the requested lengths are extracted from the file and stored in a separate structure. During the third step, the stored information about the lengths of the curve at each step is used for visualisation and analysis. The general workflow of a MATLAB® script diagram is depicted in Figure 6.7.

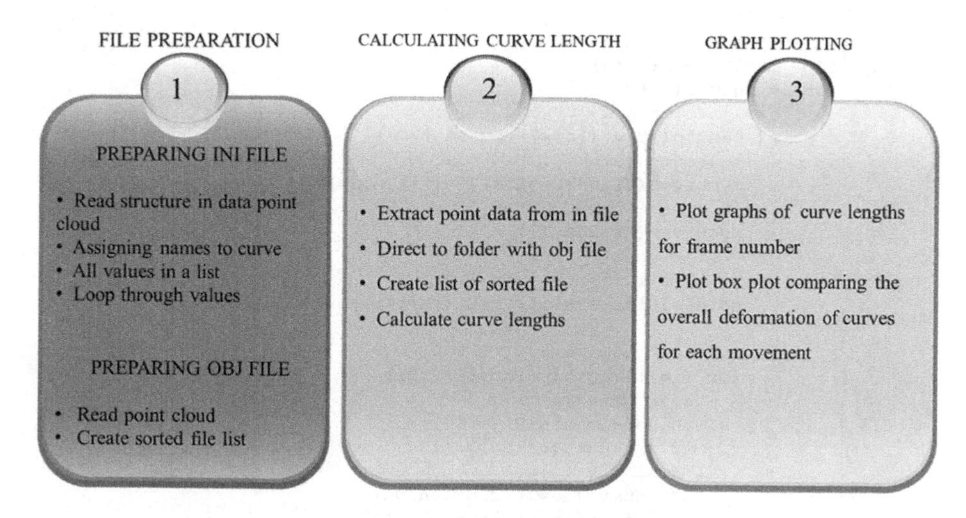

FIGURE 6.7 Workflow of MATLAB® script diagram.

```
1 [waistcircumference]
2 points=9257;23631;23632;9137;9068;9085;23356;22864;22857;8896;8895;
3 name=Waist Circumference
4
5 [legleftback]
6 points=19251;19254;19464;12887;12875;20664;12859;12862;20766;13116;13105;13086;
7 name=Leg Left Back
```

FIGURE 6.8A Example of ini file (the list of vertices is not complete) for path definition.

```
1  % (C) Vanda Tomanova and Yordan Kyosev 2022
2  function mycurves=prepare_structure2(filename)
3  % version 2
4  % each curve is in a separated field
5  %mycurves(1).name="asdfa"
6  %mycurves(1).points=[1 2 3 4 ]
7  %mycurves(1).xyz(:,:,iframe)=[1 2 3; 1 2 3; 3 4 5]
8  %mycurves(1).L(iframe)=[1 2 3; 1 2 3; 3 4 5]
9
10 %% prepare the structure for getting data
11 % read structure with the datapoints
12 stru = ini2struct(filename)
```

FIGURE 6.8B ini file preparation code, part one.

The points of each curve are saved internally as an ini file. This format is used in Windows system and contains information about parameters, option settings, and preferences applied to this system (Reviversoft n.d.). It has the advantage because it is human readable and editable and simpler for understanding for non-computer affine people compared to XML or JSON file formats, for instance. Figure 6.8a represents examples of ini files for definition of vertices and the curve names.

```
14  % get the field names
15  fn = fieldnames(stru);
16  for k=1:numel(fn)
17      %if( isnumeric(mystruct.(fn{k})) )
18      if( length(stru.(fn{k}).points)>0 )
19          % assign name
20          mycurves(k).name=stru.(fn{k}).name;
21          % initialie the ID's
22          mycurves(k).ids=[];
23
24          % get all values in a list
25          pointsIDlist=split(stru.(fn{k}).points,';');
26
27          % loop trough all
28          for (i=1:length(pointsIDlist))
29              % get the value
30              value=(pointsIDlist{i})
31              if ( length(value)>0 )
32                  % try to get as number
33                  numvalue=str2num(value);
34                  if (~isempty(numvalue))
35                      %append
36                      mycurves(k).ids=[mycurves(k).ids numvalue];
37                  end
38              end
39          end
40
41      end
42  end
```

FIGURE 6.9 ini file preparation code, part two.

The ini file is read in the MATLAB workspace using the ini2struct function of Andriy (MathWorks n.d.). The names of the fields are received using the built-in MATLAB function fieldnames (Line 15, Figure 6.8b). After that, with a loop through all fields, the MATLAB structure for the curves is filled (Figure 6.9).

The structure "my curves" is then filled with the coordinates of the data points, which are defined. The field name (without empty spaces) is used as curve ID (ids); the field "name" can contain empty spaces and capitals and is used to define the curve name for the plots; the xyz is a three-dimensional matrix, containing the coordinates of the points, which define the path, as a single 2D matrix (list of x,y,z values) per frame. Finally, the field "L" contains the computed curve lengths at each frame.

6.5.2 Preparation of the Main Part of the Code

The main part of the code reads the point definitions from an ini file (line 7 at Figure 6.10), and then asks the user to select the directory with the stored OBJ files (line 9, Figure 6.10). The list of the file names is sorted additionally, in order to be sure that the time sequence is read in the same order as it is scanned. For this purpose, the file names have to contain continuous numbering or time stamp information. Finally, the

```
1   % (C) Vanda Tomanova & Yordan Kyosev 2022
2   clear all
3   close all
4
5   print_point_numbers=true;
6
7   mycurves=prepare_structure('vanda_pointscleanlines.ini')
8
9   mydir = uigetdir();
10  %mydir='U:\studenten\2022_Vanda\workfield'
11
12  resultdir= [mydir '\results\'];
13
14  if ~exist(resultdir, 'dir')
15          mkdir(mydir , 'results')
16  end
17
18  % get sorted list of the files
19  listing=dir(fullfile(mydir,'*.obj'));
20  listing=natsortfiles(listing)
```

FIGURE 6.10 MATLAB® code, main part, part one.

```
26  % now process
27  for k = 1:numel(listing)   % for loop trough all FILES (frames)
28      F = fullfile(mydir,listing(k).name)
29      % get all curves from this frame
30      clear v
31      clear f
32      [mycurves, v, f]=get_data_from_obj(F, k, mycurves);
33      if k==1
34          v1=v;f1=f;
35      end
36      if k==10
37          v10=v
38          f10=f;
39      end
40
41  end
42
```

FIGURE 6.11 MATLAB® code, main part, part two.

OBJ files (frames) are looped through (Figure 6.11). After the processing, each curve is plotted on a separate figure. Afterwards, the lengths are calculated by calculating the distance between the points.

The amount of data that is being processed is large, so this is necessary in order to save on processing time. "Print point numbers" means that the points will be labelled with their assigned numbers; this command was optional: "**true**" meaning it was used and "**false**" meaning it was not used. This has shown beneficial for the point selection correction process. The curves data are taken from the ini file using the "prepare structure" script to read them; in brackets is the latest used ini file name. The OBJ files are then opened; "**mydir=uigetdir();**" asks where the OBJ files are located. A second option ("**mydir='U:\studenten\2020_Vanda**

workfield"') that leads directly to a folder was also used for trial versions and data corrections. New folder "results" was created in the chosen folder; here the final figures are saved in order to keep the original data clean. Afterwards, the script checks if the result folder exists. The OBJ file is then opened, and all OBJ files are sorted into an ordered list.

Figure 6.11 shows processing of the OBJ folder. Each curve k is looped through each file in the ordered list of the OBJ files. Each file is opened; all vertices and faces are opened along with the curves and sent for processing "**get_data_from_obj**". Data is then saved from the first to tenth steps for plotting.

Figure 6.12 shows the process of plotting each curve on a separate figure "**iframe=1: length (my curves (k).xyz(1,1,:))**". Each curve is plotted in the xyz coordinates with a chosen line width. Text labelling each point with its assigned number was printed for a curve on **one frame** of the OBJ folder "**iframe==1: length**". The graph is labelled by names given in the ini file.

Figure 6.13 shows the final plotting of one frame as a point cloud, that is used for orientation of the curves, defining the vertices, aspect ratio, and that the graph is plotted vertically. Finally, the resulting figure is saved in the result folder, which was created in the beginning. The change of curves throughout the movement is shown by plotting the length of each curve for each frame. The x axis shows the frame number, and the y axis shows the length in metres. This allows us to see the progression of deformation throughout the movement. A box plot of each movement was also created to show a comparison of deformation of curves of a movement.

```
43    % plot each curve on a separate figure
44    for k = 1:numel(mycurves)
45
46        figure(k);
47        for iframes= 1:length(mycurves(k).xyz(1,1,:))
48            plot3(mycurves(k).xyz(:,1,iframes),mycurves(k).xyz(:,2,iframes),mycurves(k).xyz(:,3,iframes)
              'LineWidth',3); hold on;
49            if (iframes==1)
50                if (print_point_numbers)
51                    for ipoint = 1:length(mycurves(k).ids(:))
52                    text(mycurves(k).xyz(:,1,iframes),mycurves(k).xyz(:,2,iframes),mycurves(k).xyz(:,3,
                      iframes),num2str(mycurves(k).ids(ipoint)),'LineWidth',1); hold on;
53                    end
54                end
55            end
56
57        end
58        title(mycurves(k).name)
```

FIGURE 6.12 MATLAB® code, main part three.

```
61            plot3(v1(:,1),v1(:,2), v1(:,3),'.'); hold on;
62            daspect([1 1 1])
63            axis off
64            view([0 90])
65
66            %figure2xhtml(mycurves(k).name)
67            savefig([resultdir mycurves(k).name])
68    end
69
```

FIGURE 6.13 MATLAB® code, main part, part three.

6.5.3 MeshLab

MeshLab was used to define the curves—point by point creating an ini file to be used in the MATLAB® script afterwards (Figure 6.14). MeshLab shows the avatar with mesh, texture and allows seeing the number that defines the vertex. However, the Meshlab software does not allow marking the already used point. This means that the process of defining the curve requires a lot of concentration. Defining point by point is a time-consuming process as it is; however, without marking the individual points, there is a high chance of error. The lack of marking presents extra difficulty when it is necessary to change the point of view. This can lead to losing the last used point. Finding the right point coordinate after is another issue. This is more relevant for defining longer curves.

The issue of following lines was solved by using an avatar with a texture that contains landmarks (Figure 6.15). By following the landmarks, it becomes easier to mark the points and follow the same curve even when changing the point of view.

Changing the viewing angle can present new difficulties (Figure 6.15). Certain areas can be visible, but since the rotation of the body is not fully flexible, it is impossible to reach certain areas. Vertices below the third marked point are hardly readable.

It is possible to select vertices directly in the Meshlab software (Figure 6.16); however, starting the selection with a face of a polygon is easier because the position of the freely selected vertices does not allow for precise knowledge of the position of the vertice.

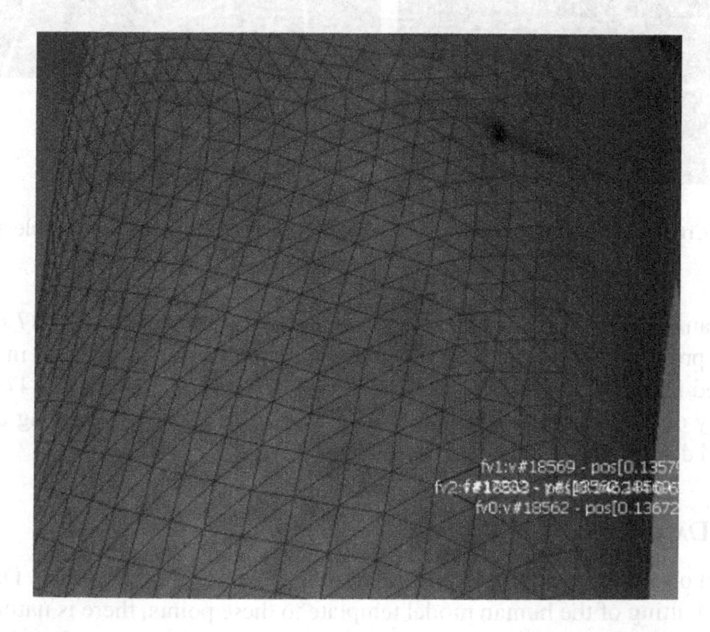

FIGURE 6.14 Change of point view in MeshLab.

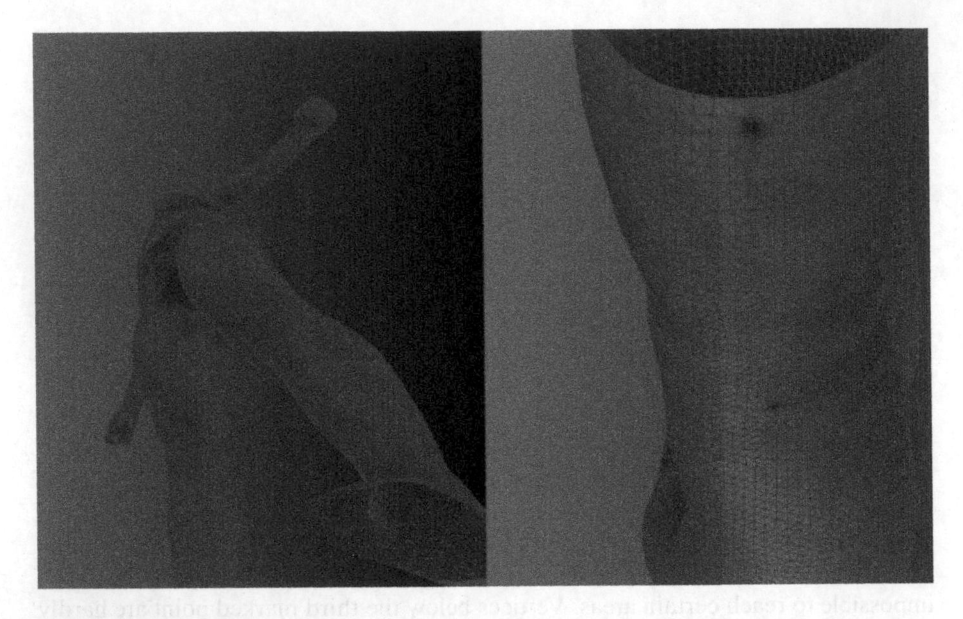

FIGURE 6.15 Change of point view in MeshLab.

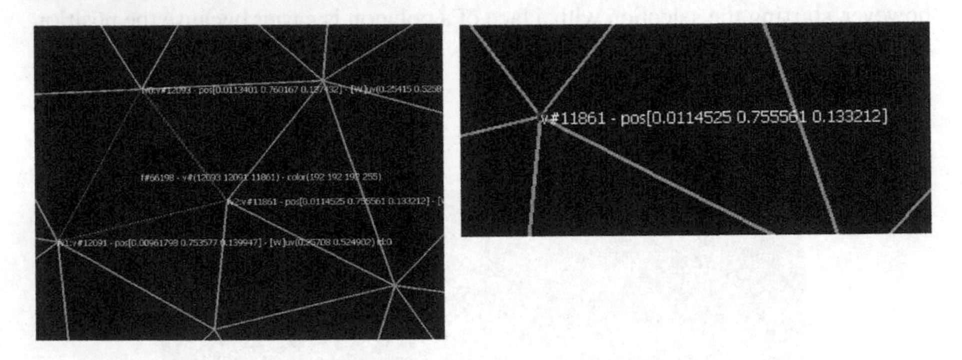

FIGURE 6.16 Vertices and points in MeshLab: (a) face selection, (b) vertex selection.

An example of marking a short line of coordinates is shown. Figure 6.17 shows the selection process in Meshlab. The following are the data points selected in Meshlab to be used in the ini file: 23288; 23244; 24446; 24389; 13793; 20873; 4732; 4735; 18222. By only marking down a certain number, the line is not creating a straight curve and does not lie directly on the surface.

6.5.4 Data Correction Process

The mesh obtained from the software is homologous, but it is too dense. During the numerical fitting of the human model template to these points, there is natural noise, where some vertices are moving around their main position in order to keep the

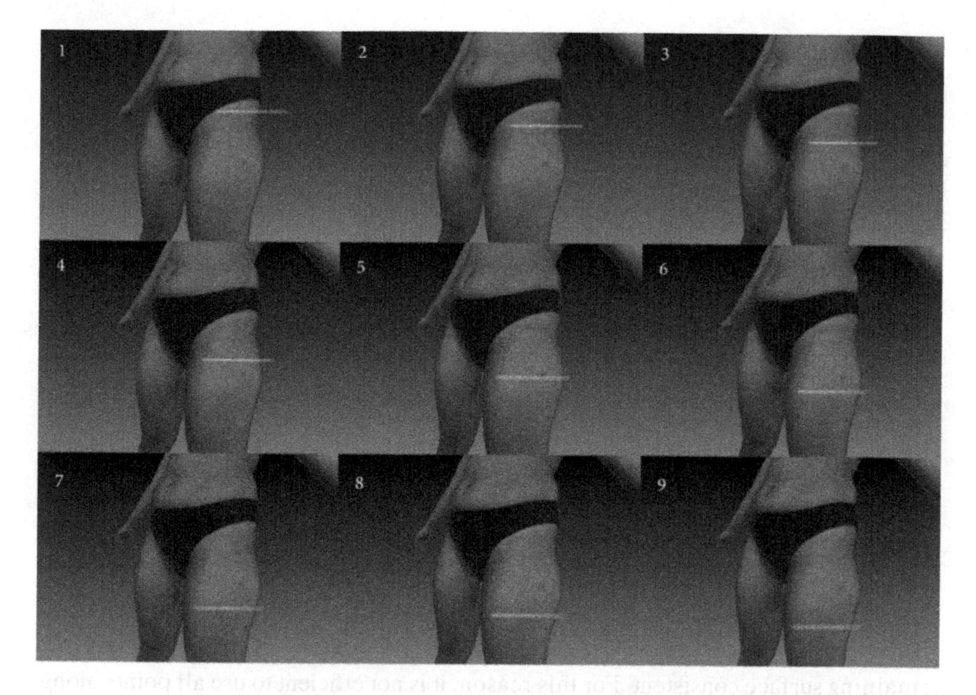

FIGURE 6.17 Shortline in MeshLab (the white text is placed close to the selected vertex).

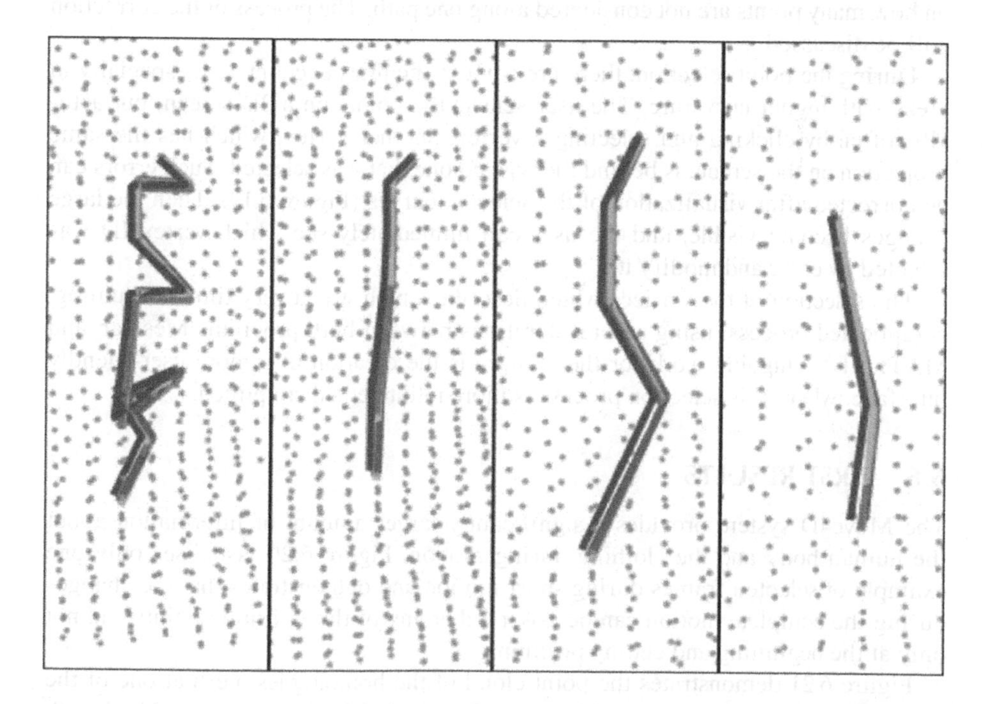

FIGURE 6.18 Zigzag MATLAB®-detail: (a) original line, (b) every second point deleted, (c) every third point deleted, (d) every fourth point deleted.

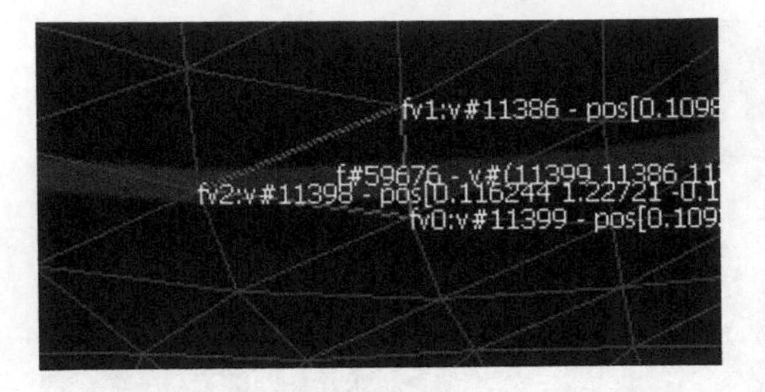

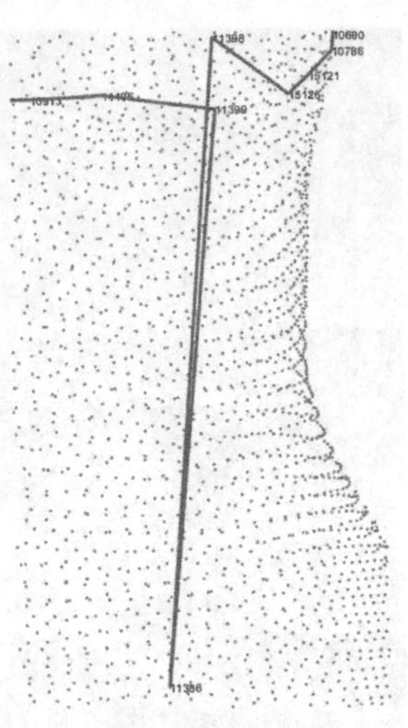

FIGURE 6.19 Selection process, issue—second hypothesis: (a) points vertices in MeshLab, (b) points vertices in MATLAB®.

remaining surface consistent. For this reason, it is not efficient to use all points along one path for its description. Figure 6.18 demonstrates some fluctuations depending on how many points are not considered along one path. The process of the correction will be discussed.

During the point selection, there are some difficulties encountered, especially in areas with higher curvature. The user selects the points on a 2D screen, but actually, often by clicking and selecting a vertex, another vertex, which has the same projection on the screen, is behind the wished one that was selected. Such errors can be corrected after visualization of the selected curves (Figure 6.19). Then the large changes become visible, and the user can immediately see which vertex ID was selected by error and modify it.

The selection of the vertices which describe a path was a very time-consuming, complicated process, using several iterations between both programs Meshlab and MATLAB®. Ongoing work for the authors is the creation of a more user-friendly interface, where this selection process is more intuitive and simplified.

6.6 FIRST RESULTS

The Move4D system provides a significantly larger amount of information about the human body and the clothing during motion. Figure 6.20 visualises only one example of selected frames during short motion and demonstrates that the changes during the complete motion can be observed at any of the transition points and not only at the beginning and ending positions.

Figure 6.21 demonstrates the point cloud of the homologues mesh at one of the frames and plotted the curve, which describes the left back leg motion. The length

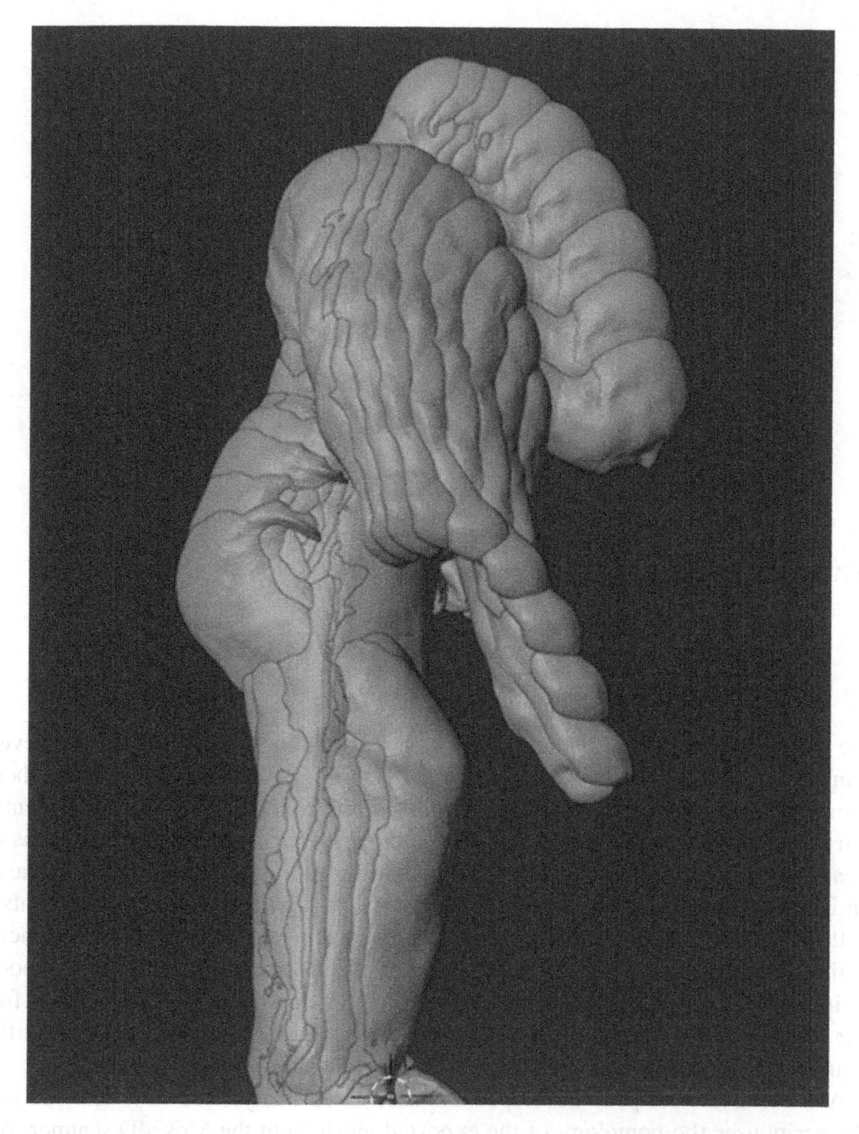

FIGURE 6.20 Sequence of motions of move scanning.

of the curve is demonstrated in Figure 6.21. The curve demonstrates that at some frames this path becomes 10 cm longer compared to its initial state. This information is important for the design of the length of the clothing parts and later for the evaluation of the deformation of the fabric during the motion.

The complete software module is now created and tested. Detailed results from its application for different motions and different types of clothing are currently in progress and will be reported during the following scientific events and papers.

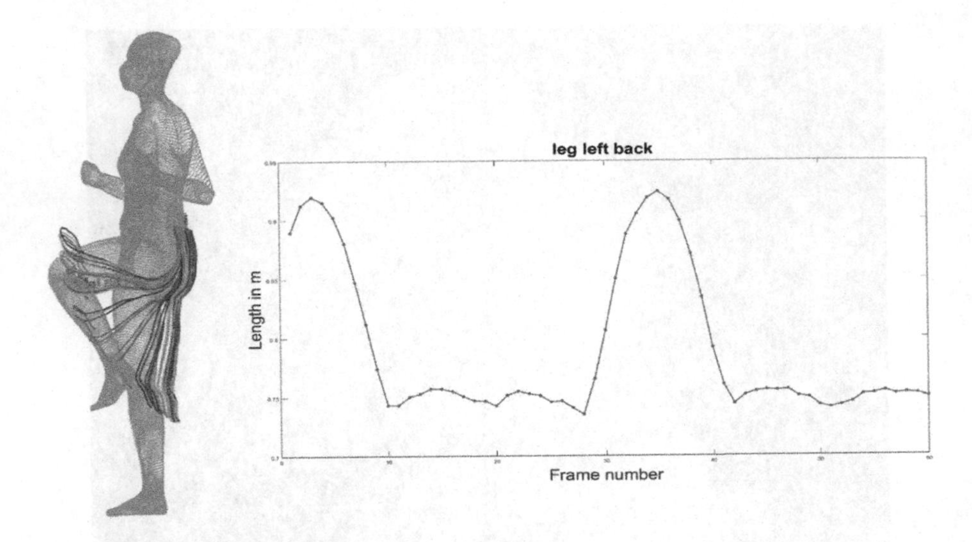

FIGURE 6.21 Sequence of motions of move scanning.

6.7 CONCLUSIONS

The digitalisation of the fashion industry is an ongoing process driven by developments in scanning and computing technology. Digital data are used everywhere in product development in order to optimise fit and functionality. The implementation of technologies such as 3D and 4D scanning systems into garment design is of great interest to consumers and producers in order to improve products in the fashion industry. Due to the high interest in digital anthropometry, this work presents a methodology of implementing a 4D scanning system provided by IBV (Valencia, Spain) for the analysis of changes in the lengths of specific paths of the human body during motion. Digital data of the human body in motion were recorded and further used for processing to extract anthropometric data in motion generated by the sequence of movements.

A MATLAB® script is developed for automated analysis of skin deformation. The script uses the homology of the exported mesh from the Move4D scanner. All steps as the point selection process through MeshLab and the main parts of the MATLAB® script are explained, including necessary corrections and hypotheses of causes of these problems. The first result of the curve evaluation demonstrates that the 4D technology combined with automatic data analysis can provide very promising results for optimisation of the clothing, providing data for the body geometry changes during motion.

REFERENCES

Alemany, Sandra, Alfredo Remon, Alfredo Ballester, Juan Vincente Durá, Beatriz Nácher, Eduardo Parrilla, and Juan Carlos González. 2022. "Data management and processing of 3D body scans." In *Digital Manufacturing Technology for Sustainable Anthropometric Apparel*, edited by Norsaadah Zakaria, 97–116. Woodhead Publishing: Sawston, Cambridge. doi.org/10.1016/B978-0-12-823969-8.00007-1.

All3DP. n.d. Accessed June 2022. https://all3dp.com/1/obj-file-format-3d-printing-cad/.

Barto, Kristijanl, David Bojanić, Tomislav Petković, and Tomislav Pribanić. 2021. "A review of body measurement using 3D scanning." *IEEE Access* 9: 67281–67301.

Bragança, Sara, Pedro Arezes, Miguel Carvalho, and and Susan A shdown. 2016. "Effects of different body postures on anthropometric measures." In *Advances in Ergonomics in Design: Advances in Intelligent Systems and Computing*, edited by Francisco Rebelo and Marcelo Soares. Cham: Springer.

Brigandt, Ingo. 2002. "Homology and the origin of correspondence." *Biology and Philosophy* 389–407. doi.org/10.1023/A:1020196124917.

De Raeve, Alexandra, Joris Cools, and Vasile S imona. 2018. "Customization business model for the clothing industryD body scanning as a valuable tool in a mass customization business model for the clothing industry." *Journal of Fashion Technology & Textile Engineering* S4: 1–6.

Giachetti, Andrea, Francesco Piscitelli, Valentina Cavedon, Chiara Milanese, and Carlo Zancanaro. 2015. "Automatic analysis of 3D scans of professional athletes." Lugano, Switzerland: 6th International Conference on 3D Body Scanning Technologies.

Gilligan, Ian. 2016. "Clothing (main entry)." In *Encyclopedia of Evolutionary Psychological Science*, edited by Todd K. Shackelford and Viviana Weekes Shackelford. Switzerland, Cham: Springer.

Hohenstein. 2019. *Hohenstein Launches Series of 4D Scanning Projects*. Hohenstein, Germany: Hohenstein Institute.

Klepser, Anke, and Simone Morlock. 2020. "4D scanning: The dynamic view on body measurements." *Communications in Development and Assembling of Textile Products*, 1 (1): 30–38.

Klpeser, Anke, Angela Mahr-Erhardt, and Simone Morlock. 2021. "Investigating fit in motion with a 4D photogrammetry scanner system." Lugano, Switzerland: 12th Int. Conf. and Exh. on 3D Body Scanning and Processing Technologies.

Kuehn, Tino, and Yordan Kyosev. 2021. "4D scanning of clothed humans: Preliminary results." 12th International Conference and Exhibition on 3D Body Scanning and Processing Technologies. Lugano, Switzerland: 3DBODY.TECH 2021. doi: 10.15221/21.25.

Lin, Huangmei, Ellen Wendt, Doudou Zhang, Jessica Boll, Jana Siegmund, Sybille Krzywiski, and Yordan Kyosev. 2021. "User-oriented product development with advanced scan solutions." 12th Int. Conference and Exhibition on 3D Body Scanning and Processing Technologies, Lugano, Switzerland.

Lindell, Eva, Hanna Tingsvik, Joel Peterson, and Li Guo. 2021. "3D body scan as anthropometric tool for infividualized prosthetic soks." *Autex Research Journal* 22 (3): 350–357.

Markets and Markets. 2022. https://www.marketsandmarkets.com/.

MathWorks. n.d. *MathWorks*. Accessed June 2022. https://www.mathworks.com/help/vision/ref/pcregistercpd.html.

Max Planck Institute for Intelligent Systems. June 2022. https://is.mpg.de/.

Medina-Inojosa, Jose, Virend K. Somers, Taiwo Ngwa, Ling Hinshaw, and Francisco Lopez-Jimenez. 2016. "Reliability of a 3D body scanner for anthropometric measurements of central obesity." *Obes Open Access* 2 (3).

Nicolau, Ana V. Ruescas, Helios De Rosario, Fermín Basso Della-Vedova, Eduardo Parrila Bernabé, M. Carmen Juan, and Juan López-Pascual. 2022. "Accuracy of a 3D temporal scanning system for gait analysis: Comparative with a marker-based photogrammetry system." *Gait & Posture* 97: 28–34. doi.org/10.1016/j.gaitpost.2022.07.001.

Owen, Richard. 1846. *Lectures on the Comparative Anatomy and Physiology of the Vertebrate Animals: Delivered at the Royal College of Surgeons of England, in 1844 and 1846. Part 1 Fishes*, Vol. 2. London, England: London, Longman, Brown, Green, and Longmans. doi.org/10.5962/bhl.title.115683.

Parker, Christopher J., Simeon Gill, and Stephen G. Hayes. 2017. "3D body scanning has suitable reliability: An anthropometric investigation for garment construction." Montreal QC, Canada: 8th International Conference and Exhibition on 3D Body Scanning and Processing Technologies.

Parrilla, Eduardo, Alfredo Ballester, Francisco Parra, and V. Ruescas Ana. 2019. "Move 4D: Accurate high-speed 3D body models in motion." 10th Int. Conference and Exhibition on 3D Body Scanning and Processing Technologies, Lugano, Switzerland, 30–32.

Pei, Jie, Linsey Griffn, Susan P. Ashdown, Jintu Fana, Bethany Juhnked, and Christopher Curry. 2021. "An exploratory study of bust measurements during running using 4D scanning technology." *International Journal of Fashion Design, Technology and Education* 14 (3): 302–313.

Pie, Jie, Linsey Griffin, Suzan P. Ashdown, and Jintu Fan. 2021. "Monitoring dynamic breast measurements obtained from 4D body scanning." *International Journal of Clothing Science and Technology* 33 (5).

Polthier, Konrad, and Markus Schmies. 2006. *Straightest geodesics on polyhedral surfaces.* New York, NY, United States: Association for Computing Machinery.

Possi, R.M. 2018. "High-performance sportswear." In *High-Performance Apparel*, edited by John McLoughlin and Tasneem Sabir. Sawston, Cambridge: Woodhead Publishing.

Quadricsurface1.docx. 2009. "Quadricsurface1.docx—Jim Lambers MAT 169 Fall Semester 2009–10 lecture 17 notes." 5–10. Retrieved May 23, 2022, from https://www.coursehero.com/file/41900754/quadricsurface1docx/.

Rajulu, Sudhakar, and Brian D. Corner. 2013. "3D surface scanning." In *The Science of Footwear*, edited by Ravindra S. Goonetilleke. London: CRC Press Taylor & Francis Group.

Reviversoft. n.d. *Reviversoft*. Accessed June 2022.

Rumbo-Rodríguez, Lorena, Miriam Sánchez-SanSegundo, Rosario Ferrer-Cascales, Nahuel García-D'Urso, Jose A. Hurtado-Sánchez, and Ana Zaragoza-Martí. 2021. "Comparison of body scanner and manual anthropometric measurements of body shape: A systematic review." *International Journal of Environmental Research and Public Health* 18 (12).

Sarkar, Debashi, and Krishanu Deyasi. 2019. "Computing the geodesic distance between two points in a polyhedral solid." *International Journal of Advanced Science and Engineering* 6 (S1): 21–24.

Sohn, Jae-Min, Sojung Lee, and Dong-Eun Kim. 2020. "An exploratory study of fit and size issues with mass customized men's jackets using 3D body scan and virtual try-on technology." *Textile Research Journal* 90 (17–18): 1906–1930. doi:10.1177/0040517520904927.

Stöcker, Björn, Daniel Baier, and Benedikt M Brand. 2021. "New insights in online fashion retail returns from a customers' perspective and their dynamics." *Journal of Business Economics* 91: 1149–1187.

Tama, Derya, and Ziynet Öndoğan. 2020. "Calculating the percentage of body measurement changes in dynamic postures in order to provide fit in skiwear." *Journal of Textile and Engineer* 27 (120): 271–282.

Textile/Clothing and Technology Corporation. n.d. *Textile/Clothing and Technology Corporation*. Accessed June 2022. https://www.tc2.com/.

Wu, Xinzhou, and Victor Kuzmichev. 2020. "A design of wetsuit based on 3D body scanning and virtual technologies." *International Journal of Clothing Science and Technology* 3 (4): 477–494.

Zhang, Doudou, Sybille Krzwinski, and Yordan Kyosev. 2021. "Possibilities for simulating clothing in motion on person-specific avatars." Lugano, Switzerland: 12th Int. Conf. and Exh. on 3D Body Scanning and Processing Technologies.

7 Smart Mirrors
Augmented, but Not Yet Reality

Hilde Heim

CONTENTS

7.1 INTRODUCTION

Augmented reality (AR) has emerged as a key interactive technology which is increasingly being utilised in the fashion retail industry (Javornik 2014; McCormick et al. 2014; Azuma and Fernie 2003). Known as 'consumer-facing' technology, AR refers to technologies and/or devices that a consumer directly interacts with

DOI: 10.1201/9781003264958-10

and experiences while shopping in a physical store or online. Computer-generated content superimposed on real-world imagery that creates a composite view on a device is a key characteristic of augmented reality technologies (Seewald and Pfeiffer 2022; El-Shamandi Ahmed et al. 2022). Examples of AR-enabled technologies include Smart, Interactive, Virtual or Magic Mirrors and mobile apps that can apply filters to a scene or a person to simulate garment try-ons. Also known as Virtual Fit technologies, these tools allow virtual garments to be superimposed on the user's 'reflection' in the 'mirror' (computer screens), mapping a realistic, 3D representation of an apparel product over the customer's body using a camera-equipped device (Saakes et al. 2016). Customers are able to try on products in a range of colours and styles without physically undressing. In this way, retailers can use AR to bridge the gap between the physical and online fashion retail experiences, turn any area into a fitting room, and provide a convenient solution to consumers' sizing issues (3DLook 2021). However, despite their apparent advantages, few smart mirrors are to be found in stores. This chapter focuses on AR-enabled smart mirror technologies, questioning their lack of adoption by fashion brands.

Virtual try-on technologies or smart mirrors (SMs) emerged in the early 2000s. The Japanese fast fashion brand Uniqlo first introduced its SM technology to consumers in 2012 in San Francisco (Holition 2018). The same year, initiated by shopping centre marketing teams, SMs appeared in malls around the world. Soon after, Neiman Marcus installed the 'memory mirror', which displayed customers' outfits in 360°, in a variety of colours, and allowed customers to share images and videos via email, social media, or with sales staff for additional recommendations (Marian 2015). Thus, consumers were able to select garments based on their needs and preferences and gain a realistic understanding of how the product may look on them. Since their launch, virtual try-on technologies have been piloted by retailers such as Nordstrom, Off White, Ted Baker, Kate Spade, Timberland, Lands' End Urban Outfitters, Rebecca Minkoff, Neiman Marcus, H&M, Zara, Ralph Lauren, Lacoste, and Burberry (CBinsights 2022; E.C. 2021). Yet, despite the promise of easy try-ons (Malik 2021), novel immersive experiences (Javornik et al. 2016), and potential to revolutionise the shopping experience, few SMs have been established in retail settings a decade later. Relatedly, despite initiatives by Snapchat and Spotify, device-based virtual try-on apps for apparel are also still very limited. In investigating the lack of universal adoption, this study first looks at the fashion retail landscape that has been evolving towards increased digital integration.

7.2 THE RETAIL LANDSCAPE

Physical retail store traffic has been declining annually (Briedis et al. 2020). In an effort to entice customers back into stores, retailers are turning to technology in the hope of providing a more engaging and innovative experience. According to several studies (Boardman et al. 2020; Fernandes and Morais 2021), AR has the ability to alter traditional shopping patterns and customers' attitudes toward purchasing fashion apparel. Indeed, according to Piotrowicz and Cuthbertson (2014), these technological advancements have increased the purchasing power and control of consumers. Fashion consumers' hunger for technology-driven experiences is growing in tandem with their usage of e-commerce (3DLook 2021; McCormick et al. 2014). As a result, the fashion retail sector is expected to be significantly impacted by technology, according to Briedis et al. (2020). According to market research (Williams 2019), the retail sector was estimated to spend $1.5 billion in 2020 on AR and virtual reality technology. The AR market is expected to reach $198 billion by 2025 (Alsop 2019). According to PWC (2020), immersive technologies are expected to enhance global GDP by $1.5 trillion by 2030. The global SM market is expected to grow from 10% in 2018 to 15% in 2023, reaching a total value of $1.2 billion, propelled largely by the retail industry (Malik 2021).

According to Porter and Heppelmann (2017), individuals can better analyse and feel connected with businesses when interacting with products that feature AR experiences. Meanwhile, Meta (2022) claims brands with an AR experience are 40% more likely to be positively evaluated, and customers who regard AR as a social activity are 20% more inclined to acquire products from the company. AR adoption is keeping pace with the rise in mobile usage: by 2025, more than 60% of social/communication apps users will be regular AR users. According to a Deloitte report commissioned by Snap Inc. (PRnewswire 2021; Snap 2021), 68% of Snapchatters of all generations utilise AR to have fun, with the majority learning about it through social media and communications apps (for example, by applying filters or stickers over images of their faces). Although AR is commonly thought of as amusement, 76% of people expect and want to utilise it as a real 'tool' in their daily lives. Shopify research found that the use of their 3D and AR app can reduce returns by up to 40% (Strapagiel 2022). According to Littledata (2022), the average conversion rate for online fashion businesses is approximately 1.8%, compared to an estimated 20%–40% for brick-and-mortar stores. Interacting with AR-enabled products results in a 94% better conversion rate, according to Kim et al. (2017).

7.3 ADVANTAGES FOR THE CONSUMER: MORE CHOICE, MORE CONVENIENCE

Millennials and younger consumers who are characterised as technologically confident (Moreno et al. 2017; Geraci and Nagy 2004; Kim et al. 2021) expect to be interested and actively engaged, and have their attention captured with new technologies. Furthermore, AR technologies, such as SMs, enable time-poor consumers to make decisions more quickly, without having to wait in line for changing rooms and/or try on different colours or sizes of the same garment (Miell 2018). Smart mirrors may improve the customer experience by reducing fitting room wait times and potentially increasing customer satisfaction, as well as reducing stress. In the words of one customer:

> I think that is a great idea! I hate to try on clothes in the stores. It takes forever to undress in each store, particularly in winter, when you are wearing multiple layers. You mess up your hair and your make up. And I always manage to leave something behind

> in the dressing room. . . . This way, you can at least narrow down your choices and try on only those you feel have the highest potential to look good on you.
>
> (EWA USA, 2021, quoted in Andrews, 2012)

The pandemic, in step with technology development, has led to several changes in the way consumers shop. The formerly separate activities of browsing and online researching (McCormick et al. 2014) have blurred and become a shopping 'skill' in their own right. Technology has served the development of this skill (Verma and Jain 2015). Customers are also more demanding of immediate gratification and convenience (McCormick et al. 2014).

7.3.1 The Technology Experiences

Although technology may offer more immediate gratification and convenience, it is not without its drawbacks. A significant difficulty with modern technology is the sensory and emotional value received from its use. The sensory dimension, in particular, will become increasingly important in the AR-related technological environment, despite the fact that it was not fully taken into account in early models. According to studies by Tueanrat et al. (2021), the sensory connection and immersive capabilities afforded by integrating the human body with a gadget (i.e., technological embodiment) contributes to the establishment of greater consumer emotional bonds. However, the tactile visceral experience is still something people want. The customer is already in the store and can try on garments in real life—viewing a reflection rather than a projection of themselves—arguably a better representation of their physical selves. Unlike a mirror which reflects a life size (and inverted) image of the viewer, the smart mirror relies on the reproduction of an image taken through a camera lens. The camera 'sees' things differently (Mistry 2021). Arguably this changes the perception of the viewer—and how they perceive themselves in the product. It may also change how they might perceive others seeing them in the product (El-Shamandi Ahmed et al. 2022). The projected image can also be rendered in different sizes. The reduction of an image changes and reduces flaws. The image that the viewer sees is open to its own interpretation. This is a distinct and emerging consumer response phenomenon already recognised for some time (El-Shamandi Ahmed et al. 2022; Kim et al. 2017). The projection of the self in images reproduced on screens aligns with a certain sense of eminence—itself bringing self-esteem. The enhancement possibilities through AR treatments (changes in skin tones, facial feature sizes, and so on) adds to the projection and perception of the restyled self, thus offering a positive experience of self while shopping.

7.3.2 Studies

Several studies have investigated whether consumers accept these new technologies and whether they are the future of the fashion retail experience. This is important knowledge to help inform brands' decisions whether or not to invest. However, little is known of the commitment by firms and in-house personnel that onboard and implement the technologies to their intended advantage. Although AR has gained increased attention and both consumers and retailers have expressed interest in its implementation, SM technology is not yet mature—or not yet truly fit for the demands of in-store deployment (Loker et al. 2008). This is partly because the technology has grown from gaming technology (Tani and Umezu 2017; Miell 2018). According to Ramanathan et al. (2014), AR technologies are still an expensive addition (LetsNurture 2022) that may not be implemented on a widespread scale for a while.

Unlike previous studies into SMs that have focussed on the consumer standpoint (Boardman et al. 2020; Briedis et al. 2020; Fernandes and Morais 2021), this study adds to the literature by taking the perspective of the organisation on the adoption, deployment, and management of emerging SM technologies within the fashion retail environment. In answering the question: what are the opportunities and challenges for fashion brands when adopting AR-enabled virtual try-on technologies?, this chapter begins with definitions of AR-enabled try-on technologies and their advantages to the retail sector; benefits and difficulties for the retailer in onboarding and maintaining the technologies to peak performance; speculations on adoption and implementation in the future, including emerging skills demands and job roles; and finally, areas for further research.

7.4 DEFINING AR APPLICATIONS

AR, the technology behind virtual try-on applications, is part of a suite of reality-enhancing technologies that come under the umbrella of extended reality (XR). Extended reality includes augmented reality (AR) and virtual reality (VR). AR and VR are still often confused (Fernandes and Morais 2021). VR is a three-dimensional illusion that involves the use of headsets, is completely immersive, shutting out the real environment, and is entirely computer generated (Farah et al. 2019). VR is outside the scope of this study, which focuses on AR only. AR presents the viewer with a hybrid of computer-generated imagery that is usually superimposed on photographic imagery—and in the case of smart mirrors, images of the customer. AR-enabled try-on applications are also confused or identified interchangeably. Applications range from virtual closets to digital clothing and include virtual showrooms, virtual fitting rooms, mobile scanning, face tracking, digital clothing, and interactive and digital display. Try-on applications are offered by tech providers such as Zugara, 3DLook, Nettelo, and Atlatl and can be categorised by function and location. AR-enabled virtual try-on smart mirror technology sits between the two poles of the AR-enabled garment simulation spectrum—between virtual closets and digital clothing. The following section discusses the various try-on technologies and associated applications.

7.4.1 Virtual Closets

A virtual closet is an application that offers an automated wardrobe styling function. The user photographs or collects similar images of all the items in their closet and uploads to a virtual closet app. Apps and websites like GlamOutfit, Closet Love, MyCloset Cladwell, and Pureple use artificial intelligence (AI) to filter and configure outfit combinations for a given occasion from an existing wardrobe, and/or make suggestions for purchase to enhance a look. They generally do not include images of the user. They do familiarise or 'gamify' wardrobe piece selection.

7.4.2 Virtual Showroom

Customers can shop in a digital environment within so-called virtual showrooms. Users can tour the environment and interact with objects as if they were in a physical store by moving their smartphone around the room. The AR Virtual Closet launched by the US store Kohl's allowed customers to explore a range of virtual try-on apparel, such as jeans, t-shirts, and jackets, and mix-and-match garments to create new combinations, all using Snapchat (McDowell 2016). The virtual showroom is increasingly found in metaverse applications.

7.4.3 Virtual Fitting Room and Body Scanners

A development of the virtual showroom, the virtual fitting room involves the integration of a customer's avatar. Through the use of a body scanner, a user's measurements are taken and reproduced digitally to create a replica of their bodies (Reid et al. 2020; Vignali et al. 2019). Body scanning usually requires a booth or static equipment such as offered by Styku (2021). Currently, avatars are not interchangeable—nor can they be exported to other programmes—but this limitation is likely to be temporary. Although arguably more accurate than other virtual try-on applications, the virtual fitting room's body scanner requirement is often prohibitive in price, bulky, and limiting in use—therefore, it's usually only installed by fashion manufacturers rather than acquired by a retailer for in-store use. Furthermore, Reid et al. (2020) found that using booths for individual customers raises concerns about body image and privacy—not to mention some inaccuracies in rendering—for example, in the angles where limbs join the torso. While body scanning booths are finding a new lease of life in the fitness industry (Li et al. 2022), their development is evolving further in garment fit, customisation, and product development rather than retailing and end consumer use.

7.4.4 Mobile Scanning

Body scanning booths have also lost their appeal for the retail consumer now that smartphones can collect body measurements with a few images. Projects such as by 3DLook, Puccto, and Nettelo provide solutions with smartphones that allow users to capture a body scan without the need for an additional attachment or specific gadget or booth (Fernandes and Morais 2021). Customers can use a camera-equipped device to try on a product. Nettelo (2021) is a body scanning app for Android and iOS that uses both the front and rear cameras on smart devices. Privacy concerns, as well as UX (user experience) and UI (user interface) issues, are the most common challenges encountered (Fernandes and Morais 2021).

Mobile scanning can be used in a number of other ways as well. Customers can hold their smartphone up to a 2D photograph of a garment and a lifelike 3D 'augmentation' appears superimposed on the garment. The 3D augmentation may also be animated—similar to a Snapchat filter on a caller's face. This is not a try-on application but provides an extended reality (XR) impression of the garment—including augmentations such as flowing in the breeze or changing colour. In applications incorporating both front and back cameras, jewellery such as watches and rings can be tried on over an image of the customer's wrist or hand (Lee 2022). Customer reviews are positive and suggest the customer's path to purchase is facilitated through their hybrid experience of amusement and enticement (Lee 2022).

7.4.5 Face Tracking

The rise of video conferencing has led to the speedier development of online filters in the business context—rather than purely for amusement. Filters help change lighting, skin tone, and facial features through digital makeup effects. The beauty industry has been quick to recognise the advantages and potential for improved sales. Companies like MAC are increasingly experimenting with this AR technology. The brand has started using SMs in its stores in collaboration with face-tracking technology provider ModiFace, which allows shoppers to try on different colours of lipstick, eyeshadow, and blush. The Perch (2022) Covergirl case study shows their platform engaged 25% of in-store shoppers, captured emails from 20% of those shoppers, and drove a 40% category lift. Cosmetics firm Charlotte Tilbury worked with Holition, a developer of augmented retail solutions, to put AR-enabled mirrors in their store. Their mirror scans the image of the customer's face and then applies a selection of the brand's looks on the customers face in the 'mirror' (Sheehan 2018). An at-home version of try-on software for the beauty industry, the MakeupGenius app from L'Oréal Paris, has been downloaded more than a million times on Google Play. It allows customers to virtually apply cosmetics to photos of their faces. Users can blend or mix and match different cosmetics to achieve their desired cosmetic look. In recent years, Saks Fifth Avenue's beauty floor and Sephora have both introduced smart mirror hardware that allow customers to virtually test cosmetics (Hirschfeld 2020). Sephora's app works a little differently to the L'Oréal application in that users must first upload a photograph of themselves on which they can apply cosmetics (Sheehan 2018). The app received 1.6 million visits and 45 million 'try-ons' in the first eight weeks. Sephora and ModiFace also combined a Nail Play station with a Virtual Make Up Wall and a smart mirror to simulate the effects of makeup, skincare, and teeth whitening goods to provide consumers a more realistic try-before-you-buy purchasing experience.

7.4.6 Digital Clothing

At the other end of the AR spectrum, so-called digital fashion is the 3D digital representation of a fashion item—not merely a photograph or 2D illustration. Using software that originates from the gaming industry, such as ZBrush, Substance, or Blender—or from digitised patternmaking or 3D prototyping software, such as Browzewear, Optitex, VStiticher, or CLO3D—an accurate garment design can be rendered in 3D. Tech companies offering digital garments include Rtfkt, The Fabricant, The Dematerialised, and DressX. Customers can send a photo of themselves to DressX, who will place a selected item of digital clothing over their bodies thus combining AR techniques and realistic photographs. Digital garments can be sold or traded as a digital asset online—commonly saved as NFTs (non-fungible tokens).

7.4.7 Interactive Digital Display

3DLOOK's YourFit, is a two-in-one AR clothing try-on and size recommendation solution. To create a photorealistic, digital replica of the client, the system recognises particular body parts (head, neck, shoulders, forearm, and ankle) and collects elements of the user's look, such as haircut and skin tone. The system that recognises body parts comes from the gaming industry's Kinect software. Kinect is a motion sensor add-on for the Xbox 360 gaming console from Microsoft. The body part recognition Kinect software is at the core of the SM technology. Timberland used Kinect technology to create a virtual fitting room. This fitting room was transformed into one of the main window displays, which was a calculated move to increase foot traffic. Shoppers would stand in front of a camera, and a virtual version of themselves would appear on a life-size screen. They could then try on several items without having to enter the store, let alone look for their size and go through the fitting room experience (Sheehan 2018). Another clothing firm that has provided clients with mobile app-driven experiences is American Apparel, using in-store signage and displays to drive engagement. A shopper would open the app and scan a picture of signage. The app would then display product information, such as user reviews, colour options, and pricing.

7.4.8 Smart Mirrors

Smart mirrors engage shoppers through motion recognition via a Kinect (or comparable) sensor and/or touch screen (nicknamed 'mirror') to market various products through visuals or animations (Fernandes and Morais 2021; Kim et al. 2017). The gadget has a natural user interface (NUI) that allows users to interact intuitively and without the use of a controller or other intermediary device. Face and voice recognition are used by the Kinect system to identify individual gamers. A player's skeletal picture is created using a depth camera that 'sees' in 3-D. The system can interpret spoken orders thanks to speech recognition software (natural language processing), and it can follow player movements thanks to gesture recognition software (Lowensohn 2011). Even though Kinect was designed for gaming, it has been used in real-world applications such as digital signage, education, health IT, and virtual shopping. Recent updates in figure recognition technology, including markerless body tracking (which detects body motion without physical markers), body segmentation (which separates figures from the environment), and pose estimation (which predicts a figure's location), are among the rapidly improving developments of the technology. A true virtual fitting room allows the user to duplicate the experience of trying on clothing, regardless of the context, without the use of an avatar, but rather through the user's 'reflection' on a screen, with live-rendered 3D animation (Fernandes and Morais 2021).

Coupled with virtual showroom technology, the ideal SM experience uses RFID (Radio Frequency Identification) whereby the user holds an article of clothing in front of the mirror and the image is scanned and saved—or captured and recognised, then retrieved from its data bank. The customer is then scanned by the mirror, which creates a virtual reproduction of their body wearing the scanned item of clothing. The virtual model in the mirror moves in real-time to display 360° views. It displays the outfit and uses motion-rendering technology to allow the user to experience the looks in different motions and light exposures. Customers can use the mirror display to request clothing in various sizes, colours, or even products that are frequently purchased together. Customers may envision alternative outfits and options. However, currently smart mirrors superimpose garments in a manner that is extremely artificial and does not replicate the sensation of wearing a real garment.

7.4.9 Retailer Adoption

The introduction of technological innovations, such as AR, can create a strong bond between customers and brands, as the association with or 'halo' effect of technology fosters excitement and a desire to explore new boundaries. This implies that retailers must invest more in technology if they wish to attract customers and increase sales in the future. Consequently, the question arises as to why all retailers do not invest in new technologies, given the potential to increase customer satisfaction and loyalty. This research investigates how willing brands are to adopt SMs, and, importantly once acquired, how brands are to adapt (including the installation and maintenance of the SM capability) in order to provide the expected customer experience.

7.5 THEORETICAL FRAMEWORK

Although there have been numerous studies of AR use from the consumer perspective (Boardman et al. 2020), this study addresses the challenges and opportunities of implementing AR technologies from the brand's perspective, and investigates the implications for the future of technology integration. In order to analyse the path to technology adoption, it is useful to refer to some of the theories that interpret the process of digital transformation. The attitude-acceptance gap of technological innovations has been a long-standing research topic. The Technology Adoption Model (TAM) developed by Davis et al. (1989) in the 1980s discusses the opposition to end-user systems as a prevalent issue in the process of digital transformation. TAM focuses on two key aspects, namely perceived usefulness and perceived ease of use, to determine an individual's intention to use new technological innovations. The TAM model provides insight into why users may accept or reject the use of a particular technology depending on its ease of use. In the 1990s, Rogers (2003) contributed the Diffusion of Innovations (DoI) theory, which explains user reluctance in technology adoption and lays out a curve of phases from 'early adopter' to 'laggards'. The marketing of high-tech and innovative products requires a leap across 'The Chasm' between early adopters and early majority acceptance, once 'The Tipping Point' between the two has been reached. If the tipping point is not attained, the product is unlikely to see full acceptance.

The Adoption of Innovations curve and the Technology Adoption Model (TAM) show that new technologies that are introduced to the market require some time to take hold and then eventually either wane in popularity or remain embedded depending on their perceived ease of use. Prior to acceptance, the introduction of a new innovation usually first appears in the form of media hype—the technology is often not yet mature at these early stages. Lag time sets in while the emerging technology is piloted or tested in the market and its flaws are discovered and addressed. This cycle may rise and fall several times before the innovation experiences full acceptance. This study observes that AR applications such as Smart Mirrors have also been subject to what Gartner (Gartner 2019) refers to as the hype-cycle phenomenon. Stone (2015) notes that in his 30 plus years in the extended reality (XR) industry, he has seen the sector go through a seven- to eight-year 'hype and reality' cycle. He sees:

> roughly 7 year cycles of waning and waxing depending on a myriad of things, such as hype, failure of tech, failure of tech companies to deliver, start-ups with bright ideas coming and (mostly) going, adoption and rejection in different sectors.
>
> (Stone 2015, quoted in Orengo, 2022)

This study observes that try-on technology has not yet passed the early adopter phase. The reasons are manifold, according to Fernandes and Morais (2021), including the costs of development and installation, neural networks still undergoing machine learning, lack of tech know-how in fashion companies, the lack of competitors in the market to drive adoption, consumers without the latest devices, and the perceived demand among consumers. However the literature suggests that the SM is a stimulating digital addition to the shopping experience, adding to the omnichannel palette (McCormick et al. 2014), and that retailers are feeling increasingly compelled to adopt.

Although technology adoption theories grew out of the pre and early eras of Web 1.0 and Web 2.0 and indeed much of the reasoning still stands, enterprises today find themselves in a position of having some knowledge and experience of software applications—but are now faced with the even more complex programmes of Web 3.0 technologies—and their integration into existing ('legacy') systems. Whereases the TAM and DoI models focus on user acceptance, according to Hoque et al. (2021), the literature lacks investigations of technology—organisation—environment and institutional theories for the adoption of technology in industry. Therefore, an updated version of a third theory, namely the Technology Readiness Index (TRI), is applied here to augment the earlier theories and analyse transformational change. Specifically, the Blut and Wang (2020) variation of the TRI is useful here. The purpose of the TRI is to measure participants' inclination to adopt and utilise cutting-edge technologies. The TRI postulates that the firm now requires a number of measures on the path to technology adoption. These may be incremental and multifaceted, and more complex requirements than in the previous TAM and DoI eras. The Blut and Wang (2020) TRI model considers four dimensions that collectively describe technology usage: innovativeness, optimism, insecurity, and discomfort. This new complexity justifies the application of the TRI theory which suggests new understandings and alignment on process and information workflows, standardisation of terminology, and governance around disclosure as well as cultural and organisational adjustments, including appropriately skilled personnel, are required within the firm. Considering the eCommerce platform reports provided by Shopify, Deloitte, and other technology providers, the findings suggest retailers may well assume that try-on technologies will lead to improved customer journeys, better brand loyalty, and increased purchases. Therefore, examining technology adoption from the retailer's standpoint warrants investigation.

7.6 METHODOLOGY

This study sets out by investigating AR-enabled try-on technologies—their delivery on promise; their uptake, implementation, and adoption by retailers; and the impact on the customer's path-to-purchase. The study undertakes an analysis of media reports and reviews that appear on brands' websites or are reported in the professional fashion and/or technology press. This textual review, which provides vignettes of use cases, is supplemented with empirical data gathered from one operator in an academic setting installing and implementing the technology. Rather than using the SM with prior fitted systems and generic assets (plug and play), the operator has installed all the necessary software manually. This gave the operator significant background insight into the requirements behind the SM installation and thereby generated supplementary data for this study. The operator was interviewed for this study and gave short demonstrations and an outline of instructions on the various software applications involved. The data gained from engagement with the operator is analysed here through the technology adoption models discussed earlier.

7.7 AR OPPORTUNITIES AND CHALLENGES

The following section provides vignettes of use cases from brands that have adopted smart mirrors. The vignettes are analysed by applying the technology adoption frameworks discussed earlier, specifically applying TRI's four dimensions of innovativeness, optimism, insecurity, and discomfort. The reports are followed by empirical data on SM installation and use.

7.8 ADVANTAGES FOR THE RETAILER: AR AS A PROMOTIONAL TOOL

The digital environment in retail establishments can be a powerful tool for attracting customers to the store. Advantages include improving customer–associate connection by expediting the fitting room procedure and delivering a streamlined experience, theft protection (in the case of RFID-tagged items), and inventory control and management. Data collecting, extending loyalty programmes, sales tracking, and the social media sharing that occurs with each use; new ways to include product information; recommendations; customer service; and social media are all key considerations.

Several big retailers appear to have understood this phenomenon. For example, Burberry's flagship shop in London, which translates the brand's web experience into a physical environment, has added smart mirrors to its changing rooms and communal areas. This technology includes interactive devices that enhance the retail experience by offering specific product information such as material properties and craftsmanship, as well as displaying making-of and runway films of chosen products. This example demonstrates that the SM has been accepted for ease of use by the customer. However, the SM is functioning here as a life-size touch pad offering catalogue information—rather than as a try-on technology as such. The aforementioned example is the adaptation by a brand of only part of the technology, specifically the 'ease of use' part which is also not too demanding on human resources (the lack of staff with technical know-how) but still implements the technology to a lesser extent. Effectively this is an underutilisation of the technology as only the 'ease of use' part of the technology is implemented—and the too difficult to use parts are left dormant. The implications are that the technology's capabilities outweigh the brand's personnel capabilities. In this way the brand is under-delivering on the SM promise. It may seem that this is the brand's 'fault'—and indeed follows the technology adoption theories described previously that suggest that the operator is at fault rather than the inventor (Rogers 2003). It appears that little acknowledgment exists of the gap in skills and knowledge between technology provider and the interim position that the brand holds. It raises the question of whether the brand or the tech provider is responsible for the full deployment of the software's capabilities.

7.9 THE FIT AND RETURNS PROBLEM—AN OPPORTUNITY

Customers and retailers have yet to be supplied with a suitable solution to the well-known e-commerce challenge of fit (Fernandes and Morais 2021). According to McKinsey, 70% of fashion returns are due to poor fit or style (Briedis et al. 2020). Online shopping exacerbates the problem, with e-commerce return rates three times greater than in-store returns. Fit analytics firm Arnett, specialising in fitting algorithms that began with a webcam-based solution in 2010, is currently one of the most widely used solutions to the e-commerce fit problem using the comparative body type method. The business launched 'See My Fit', an inclusive AR platform that allows buyers to try on a selection from 800 dresses online. The See My Fit app displays apparel on models that are more accurate representations of specific clients. The company then created an algorithm that retailers can employ on their websites. However, the company ran into the same problem as many other solutions: the entire process was overly taxing on users—and users prefer their own bodies rather than a simulated body type. Even if companies were 'early adopters', according to the DoI model, the innovation did not breach 'the chasm' through the tipping point to the 'early majority' adoption stage. Applying TRI's dimensions, customer discomfort outweighed the innovativeness of the application. Similarly, applying the TAM model, perceived ease of use weakened the application's perceived usefulness.

7.9.1 The Material Problem—A Challenge

In the footwear market, Nike, Converse, Gucci, Reebok, and Allbirds have added sneaker try-on capabilities to their online storefronts thanks to solutions like Wanna Kicks. Customers may now visualise how a shoe will look on their foot using a smartphone device. Converse created an iPhone AR app that allows customers to visually try on trainers by pointing their phone at their legs. Customers can also place orders straight through the app without having to leave their homes. Lacoste has released an AR mobile app called 'LCST Augmented Reality' that allows customers to visually try on new shoes utilising displays in Lacoste stores (Hirschfeld 2020). More than 30,000 people used the app to interact with 3D items (E.C. 2021). Some companies have gone so far as to include AR into the product itself, improving not only the purchasing experience but also the overall brand experience. Adidas, for example, released a range of sneakers that enabled users to use AR at home. Customers would take the sneakers home after purchasing them and hold them up to their computer's webcam so that the hidden code on the tongue could be read. Customers would then be immersed in a virtual environment that they could explore using their sneaker as a controller (Sheehan 2018). As mentioned earlier, hard, small, and/or fixed objects are easier to apply to AR technologies. Here innovativeness and optimism override insecurity and discomfort. Indeed, there is very little insecurity and discomfort to experience in the accessories AR experience. Arguably, the technology will be rapidly adopted for this market. In another example, Specsavers has integrated an AR component into its website and mobile app to assist customers in choosing the correct pair of glasses. Customers might digitally try on glasses by superimposing the frames on their faces using the technology. AR in particular, as seen in Specsaver's glasses try-on mobile app and website functionality, could help customers have a better omnichannel experience by bridging the gap between channels and limiting their downsides (Boardman et al. 2020).

Clothing try-on apps have proven to be more elusive and difficult to implement than their cosmetics, footwear, and accessory cousins (McDowell 2021). Comfortable fit and size, product 'touch and feel', zero delivery lead time, and guidance and up-selling from sales professionals have all been advantages of physical stores (McCormick et al. 2014). The most significant problem for AR is to correctly detect and represent the flexibility and drape of fabric on the body—and for that matter, the moving body. This means that other items such as makeup and accessories (eyewear and footwear) are more suitable for the current state of try-on software because they are rigid objects. According to Fernandes and Morais (2021), the apps for clothing are severely limited because the clothing items are presented as two-dimensional images, and they require users to adjust the silhouette to the Kinect sensor screen image of themselves, which they frequently fail to do, often crashing the software. Furthermore, enlarging an image of a garment does not follow patternmaking principles and will give a false impression of the size and fit. The bulk of visualisations on smart mirrors superimpose garments in an extremely artificial way. Superimposing images of garments is not a satisfactory solution for customers.

Despite the hype, form-fitting clothing that follows the rules of physics and adheres to a variety of body types are still on the horizon. Cloth simulation is a problem. Part of the issue, according to Vodolazov (2021), is a lack of optimum 'tracking accuracy', which means the clothing appears to be laid on top of the wearer rather than fitted around them (McDowell 2021). The Fashion Innovation Agency recently collaborated with Digital Domain, a visual effects agency, on a project to recreate hyper-realistic textiles using machine learning (McDowell 2021). This is less of a problem in some SM technology when a customer tries on an actual garment in store and then can change the colour variations in the mirror—in which the software simply recolours the projections of the actual garment on the body.

7.9.2 The Technology Problem

The technology adoption models mentioned earlier focus on the user experience—but little is discussed on the installation and implementation aspect of new software–hardware packages such as SM installation. Setting up the technology is not without its challenges. When engaging with the technology—or moreover, installing the technology for use—it becomes evident that a number of software applications are required to coalesce to deliver a seamless and 'useful' experience. Herein lies one of the biggest challenges for the brand. In order to deliver an optimum experience—not merely a life-size catalogue or 'paper doll' effect—the SM technology requires the addition of at least five highly sophisticated software applications. These include (1) 3D prototype modelling software such as CLO3D, Browzewear, and or Optitex to create three-dimensional garments—that are not mere flat images of the garment; (2) the generic, built-in avatar which requires adaptation for correct garment fit; (3) the 3D garments which require conversion through a gaming software like Unity that enables interactivity; (4) a degree of HTML coding; and (5) the utilisation of the SM software to upload schedules of catalogues of garments. In this case, ease of use becomes moot, and innovativeness and optimism are significantly outweighed by the discomfort of not managing the software correctly, crashing the software, or the insecurity that the software will not work as intended.

The hardware of the SM (a screen approximately 1.8 metres × 1 m × 30 cm) is large and cumbersome—but still smaller than the square meterage of one change room. It is also portable to a degree. The cost is equivalent to a large smart TV. Therefore, it is considered affordable as a retail fixture. It does come installed with some basic software. However, to reach its full potential, a number of installation steps are required, including coding. This is an additional service that can be offered by the SM provider but comes with costs significantly greater than the hardware itself[1] (LetsNurture 2022). In the words of one operator:

> The process to make virtual try-ons happen is in many ways quite complicated. Most people that are involved in (developing) this kind of technology are not fashion people they are technical people. Normally you need quite a set of skills to make this work. Companies that sell magic mirrors will normally do the asset creation for you but obviously there is a cost in doing that.

Therefore, adoption is hindered by a combination of costs and technical know-how—thus optimism is thwarted by uncertainties.

> It is possible to create the assets yourself, but it is not straightforward. You need to learn a lot of software; so, for example there is Unity software you have to learn; the CLO3D or any 3D prototyping software is also needed.

It becomes increasingly evident that proficiency in sophisticated software programmes is non-negotiable here—however, brands may not feel this is worth the return on investment as it distracts from the core purpose of the fashion brand.

> You then need to export your file out and you need to make it fit the 3D avatar that has been designed for the smart mirror; because obviously the avatar has to be of a particular shape to work on the mirror—you can't export out any avatar. Once you get to a point where you make your 3D garment fit their avatar, you have to go through quite a long step by step process basically converting that in Unity. Then the cloth needs to be created with a texture map—or you pay for the company to convert the garments for you.

That can become very costly and time-consuming per garment. It may be feasible in the future that if a garment is already designed using virtual design software, the 3D prototype can be used for the SM. A library of assets then needs to be created, for example, women's wear collection 1, 2, 3. These can be scheduled to 'drop' as appropriate for the season. HTML coding is also required to make images of garments appear to fit on the body.

These are all time-consuming tasks requiring dedicated technicians. The SM therefore is not merely a 'plug-and-play' proposition for the store. Indeed, this may suggest new studies/models are required that address the phase between technology and installation, as well as maintenance, and the skills, mindset, and preparation required for this 'bridging' stage and or function by an organisation.

7.10 CONCLUSION

The use of AR in fashion presents several new opportunities as well as challenges for fashion retailers. AR in particular has piqued the interest of retailers because it offers a variety of solutions—from already in-use technology to more sophisticated equipment and the halo effect associated with cutting-edge technology. As online sales have increased, AR could help revitalise the high street by providing consumers with engaging, entertaining, and useful experiences. The physical retail environment will continue to play an important role in future fashion retailer strategy, with the construction of new locations that 'amaze and amuse' (RetailWeek 2022; Kim et al. 2020). Therefore, a successful long-term retail approach includes the adoption of consumer-facing digital technology that integrates physical and digital channels aligning with the retailer's positioning strategy. Smart mirrors driven by AR are becoming a part of both physical and online purchasing and are set to significantly change the way customers purchase. However, as observed by Gartner (2019), the media is enthusiastically presenting novel technologies, sometimes unrealistically building new expectations in audiences. Despite the cases cited by the media, the technology has only become ubiquitous in the mobile market—with face rather than full-body recognition. It relies on simple makeovers that include colour changes or that involve rigid artefacts such as eyewear and footwear. The fluid 3D rendering of real garments on real bodies is some time off.

Existing research on AR in the context of fashion retail and its impact on the industry as a whole is limited, but there is evidence of its potential advantages. The perceived usefulness and usability of AR technology implementations, which are mediated by the intention to use these technologies, can be used to measure their success (Davis et al. 1989). This research demonstrates that opposition to end-user systems as seen in the TAM model is not necessarily the case with SM technology—rather it is the complexity of installation and the multifaceted process required by the organisation to do so—often without sufficiently skilled personnel. This research found that the gap in skills and knowledge between innovation and implementation has not been fully acknowledged. Future research could address this gap and build understanding on how brands are to approach technology adoption. This means that a new genre of hybrid fashion/technology employee may emerge and will be in greater demand. Importantly, retailers will require a team that has the combination of both fashion and technical know-how to do so effectively.

NOTE

1. The cost of implementation can range from USD 30K–USD 50K, according to app provider Let's Nurture.

REFERENCES

3DLook. 2021. "Augmented reality and fashion: The perfect fit?" *3DLook*. Accessed December 31, 2021. https://3dlook.me/content-hub/how-to-impelement-ar-in-fashion/.

Ahmed, K.E., A. Ambika, and R. Belk. 2022. "Augmented reality magic mirror in the service sector: Experiential consumption and the self." *Journal of Service Management*. doi: 10.1108/Josm-12-2021-0484.

Alsop, Thomas. 2019. "Augmented reality (AR) market size worldwide in 2017, 2018 and 2025." In *Statista*, edited by BIS Research. Hamburg: Statista.

Andrews, Emily. 2012. "'Try before you buy' 21st century style: Now there's a mirror that tries your clothes on for you." *Mailonline* 23 (March 12): 58, Femail. www.dailymail.co.uk/femail/article-2114156/Try-buy-21st-Century-style-Now-theres-mirror-tries-clothes-you.html.

Azuma, Nobukaza, and John Fernie. 2003. "Fashion in the globalized world and the role of virtual networks in intrinsic fashion design." *Journal of Fashion Marketing and Management: An International Journal* 7 (4): 413–427.

Blut, Markus, and Cheng Wang. 2020. "Technology readiness: A meta-analysis of conceptualizations of the construct and its impact on technology usage." *Journal of the Academy of Marketing Science* 48 (4): 649–669.

Boardman, Rosy., Henninger, Claudia. E., & Zhu, Ailing. (2020). Augmented reality and virtual reality: new drivers for fashion retail? In G. Vignali, L. F. Reid, D. Ryding, & C. E. Henninger (Eds.), *Technology-Driven Sustainability: Innovation in the Fashion Supply Chain* (pp. 155–172). Cham, Switzerland: Palgrave Macmillan. https://doi.org/https://doi.org/10.1007/978-3-030-15483-7_9

Briedis, Holly, Anne Kronschnabl, Alex Rodriguez, and Kelly Ungerman. 2020. "Adapting to the next normal in retail: The customer experience imperative." *McKinsey*. Accessed

July 18, 2022. www.mckinsey.com/industries/retail/our-insights/adapting-to-the-next-normal-in-retail-the-customer-experience-imperative.

CBinsights. 2022. "The future of fashion: From design to merchandising, how tech is reshaping the industry." *CB Insights*. www.cbinsights.com/research/fashion-tech-future-trends/.

Davis, Fred D., Richard P. Bagozzi, and Paul R. Warshaw. 1989. "User acceptance of computer technology: A comparison of two theoretical models." *Management Science* 35 (8): 982–1003. https://doi.org/10.1287/mnsc.35.8.982.

E.C. 2021. "LCST augmented reality retail campaign: Lacoste." *Engine Creative*, December 31. www.enginecreative.co.uk/portfolio/lacoste-lcst-augmented-reality-retail-campaign.

El-Shamandi Ahmed, Khaled, Anupama Ambika, and Russell Belk. 2022. "Augmented reality magic mirror in the service sector: Experiential consumption and the self." *Journal of Service Management* (ahead-of-print). doi: 10.1108/JOSM-12-2021-0484.

Farah, Maya F., Zahy B. Ramadan, and Dana H. Harb. 2019. "The examination of virtual reality at the intersection of consumer experience, shopping journey and physical retailing." *Journal of Retailing and Consumer Services* 48: 136–143.

Fernandes, Clara E., and Ricardo Morais. 2021. "A review on potential technological advances for fashion retail: Smart fitting rooms, augmented and virtual realities." *dObra [s]—revista da Associação Brasileira de Estudos de Pesquisas em Moda* (32): 168–186.

Gartner. 2019. "Interpreting technology hype." *Gartner*. Accessed March 03, 2020. www.gartner.com/en/research/methodologies/gartner-hype-cycle.

Geraci, John C., and Judit Nagy. 2004. "Millennials-the new media generation." *Young Consumers*.

Hirschfeld, Andy. 2020. "Augmented reality is going change how you update your wardrobe." *Observer*, December 31, 2021. https://observer.com/2020/02/augmented-reality-retailers-asos-gap-smart-mirrors-mobile-apps/.

Holition. 2018. "Uniqlo world's first magic mirror." *Holition*. Accessed July 18, 2022. https://holition.com/work/uniqlo-world-s-first-magic-mirror.

Hoque, Md Aynul, Rajah Rasiah, Fumitaka Furuoka, and Sameer Kumar. 2021. "Technology adoption in the apparel industry: Insight from literature review and research directions." *Research Journal of Textile and Apparel* 25 (3): 292–307.

Javornik, Ana. 2014. "[Poster] classifications of augmented reality uses in marketing." 2014 IEEE International Symposium on Mixed and Augmented Reality-Media, Art, Social Science, Humanities and Design (ISMAR-MASH'D).

Javornik, Ana, Yvonne Rogers, Ana Moutinho, and Russell Freeman. 2016. *Revealing the Shopper Experience of Using a 'Magic Mirror' Augmented Reality Make-Up Application*.

Kim, Ha Youn, Yuri Lee, Erin Cho, and Yeo Jin Jung. 2020. "Digital atmosphere of fashion retail stores." *Fashion and Textiles* 7 (1): 1–17.

Kim, Hye-Young, Ji Young Lee, Jung Mee Mun, and Kim K.P. Johnson. 2017. "Consumer adoption of smart in-store technology: Assessing the predictive value of attitude versus beliefs in the technology acceptance model." *International Journal of Fashion Design, Technology and Education* 10 (1): 26–36.

Kim, Songmee, Seyoon Jang, Woojin Choi, Chorong Youn, and Yuri Lee. 2021. "Contactless service encounters among Millennials and Generation Z: The effects of Millennials and Gen Z characteristics on technology self-efficacy and preference for contactless service." *Journal of Research in Interactive Marketing* 16 (1): 82–100.

Lee, Adrianna. 2022. "Perfect corp.'s fashion push hits the wrist with AR watch try-on." *WWD*. Accessed January 29, 2023. https://wwd.com/business-news/technology/perfect-corp-fashion-ar-try-on-watches-1235025982/.

LetsNurture. 2022. "How much would it cost to develop an AR based smart mirror system for retailers?" *Let's Nurture*. Accessed July 19, 2022. www.letsnurture.com/how-much-would-it-cost-to-develop-an-ar-based-smart-mirror-system-for-retailers.html.

Li, Jiayin, Sibei Xia, Andre J. West, and Cynthia L. Istook. 2022. "Fashionable sportswear working as a body measurement collecting tool." *International Journal of Clothing Science and Technology* 34 (4): 589–604.

Littledata. 2022. "What is the average conversion rate for style and fashion websites?" *Littledata*. Accessed July 19, 2022. www.littledata.io/average/ecommerce-conversion-rate-(all-devices)/Style-and-fashion-websites.

Loker, Suzanne, Susan Ashdown, and Erica Carnrite. 2008. "Dress in the third dimension: Online interactivity and its new horizons." *Clothing and Textiles Research Journal* 26 (2): 164–176.

Lowensohn, Josh. 2011. "Timeline: A look back at Kinect's history." January 4, 2022. www.cnet.com/news/timeline-a-look-back-at-kinects-history/.

Malik, Diana Lee Rahul. 2021. The opportunity in digital fashion and avatars report. In *BoF Insights*. London.

Marian, Petah. 2015. "Fashion retailer Neiman Marcus unveils digital 'memory mirror'." *Retail Week*, July 18, 2022. www.retail-week.com/tech/fashion-retailer-neiman-marcus-unveils-digital-memory-mirror/5068142.article?authent=1.

McCormick, Helen, Jo Cartwright, Patsy Perry, Liz Barnes, Samantha Lynch, and Gemma Ball. 2014. "Fashion retailing: Past, present and future." *Textile Progress* 46 (3): 227–321.

McDowell, Maghan. 2016. "Tech and social media drive insta-fashion." *WWD*: 32–32.

McDowell, Maghan. 2021. "Why AR clothing try-on is nearly here." *Vogue Business*, January 4, 2022. www.voguebusiness.com/technology/why-ar-clothing-try-on-is-nearly-here.

Meta. 2022. "Help take your business to new dimensions with augmented reality." *Meta*. Accessed July 19, 2022. www.facebook.com/business/augmented-reality/.

Miell, Sophie L. 2018. *Enabling the Digital Fashion Consumer through Gamified Fit and Sizing Experience Technologies* (PhD thesis). The University of Manchester (United Kingdom). https://research.manchester.ac.uk/en/studentTheses/enabling-the-digital-fashion-consumer-through-gamified-fit-and-si.

Mistry, Taran. 2021. "The future is here now: A look at smart mirrors." *LinkedIn*, July 19, 2022. www.linkedin.com/pulse/future-here-now-look-smart-mirrors-taran-mistry/?trk=public_profile_article_view.

Moreno, Flor Madrigal, Jaime Gil Lafuente, Fernando Ávila Carreón, and Salvador Madrigal Moreno. 2017. "The characterization of the millennials and their buying behavior." *International Journal of Marketing Studies* 9 (5): 135–144.

Nettelo. 2021. "Democratizing 3D body scan & analysis." Accessed July 19, 2022. http://nettelo.com/.

Orengo, Joh. 2022. "Pedaling the metaverse hype cycle." *Spatial News*. https://spatial8.substack.com/p/spatial-news-007?

Perch. 2022. "Retail's leading digital shopper marketing platform for product engagement in-store." *PerchInteractive*. www.perchinteractive.com/.

Piotrowicz, Wojciech, and Richard Cuthbertson. 2014. "Toward omnichannel retailing." *International Journal of Electronic Commerce* 18 (4): 5–16.

Porter, Michael E., and James E. Heppelmann. 2017. "Why every organization needs an augmented reality strategy." *Harvard Business Revue*. https://hbr.org/2017/11/why-every-organization-needs-an-augmented-reality-strategy.

PRnewswire. 2021. "Deloitte digital and snap Inc. report reveals the rich, untapped future of augmented reality for customer experience." *Cision*, July 19, 2022. www.prnewswire.com/news-releases/deloitte-digital-and-snap-inc-report-reveals-the-rich-untapped-future-of-augmented-reality-for-customer-experience-301290445.html.

PWC. 2020. *Seeing Is Believing: How Virtual Reality and Augmented Reality Are Transforming Business and the Economy*. London: Price Waterhouse Coopers. Accessed January 23,

2023. https://www.pwc.com/gx/en/technology/publications/assets/how-virtual-reality-and-augmented-reality.pdf.

Ramanathan, Ramakrishnan, Usha Ramanathan, and Lok Wan Lorraine Ko. 2014. “Adoption of RFID technologies in UK logistics: Moderating roles of size, barcode experience and government support.” *Expert Systems with Applications* 41 (1): 230–236.

Reid, Louise. F., Vignali, Gianpaolo., Barker, Katharine., Chrimes, Courtney., & Vieira, Rachel. (2020). Three-dimensional Body Scanning in Sustainable Product Development: An exploration of the use of body scanning in the production and consumption of female apparel. In G. Vignali, L. F. Reid, D. Ryding, & C. E. Henninger (Eds.), *Technology-Driven Sustainability: Innovation in the Fashion Supply Chain* (pp. 173–194). Cham, Switzerland: Palgrave Macmillan. https://doi.org/10.1007/978-3-030-15483-7_10

RetailWeek. 2022. “Retail’s shapeshifters: Delighting customers by extending the brand experience.” *Ascential Information Services.* Accessed July 20, 2022. www.retail-week.com/tech/retails-shapeshifters-delighting-customers-by-extending-the-brand-experience/7040562.article?authent=1.

Rogers, Everett M. 2003. *Diffusion of Innovations*, 5th ed. New York: Free Press.

Saakes, Daniel, Hui-Shyong Yeo, Seung-Tak Noh, Gyeol Han, and Woontack Woo. 2016. “Mirror mirror: An on-body t-shirt design system.” Proceedings of the 2016 CHI Conference on Human Factors in Computing Systems.

Seewald, Alexander K., and Alexander Pfeiffer. 2022. “Magic mirror: I and my avatar-a versatile augmented reality installation controlled by hand gestures.” ICAART (3).

Sheehan, Alexandra. 2018. “How these retailers use augmented reality to enhance the customer experience.” December 31, 2021. www.shopify.co.uk/retail/how-these-retailers-are-using-augmented-reality-to-enhance-the-customer-experience.

Snap. 2021. The Next Inflection Point: More Than 100 Million Consumers Are Shopping with AR. Online.

Stone, Robert J. 2015. *Virtual & Augmented Reality Technologies for Applications in Cultural Heritage: A Human Factors Perspective.* ISSN.

Strapagiel, Lauren. 2022. “The ROI on AR: How augmented reality is boosting ecommerce sales.” *Shopify.* Accessed January 29, 2023. https://www.shopify.com/uk/blog/ar-shopping.

Styku. 2021. “3D body scanning makes visualising progress simple.” *Styku.* Accessed July 19, 2022. www.styku.com.

Tani, Misaki, and Nobuyuki Umezu. 2017. “Prototype implementation of mirror with built-in display.” Thirteenth International Conference on Quality Control by Artificial Vision 2017.

Tueanrat, Yanika, Savvas Papagiannidis, and Eleftherios Alamanos. 2021. “A conceptual framework of the antecedents of customer journey satisfaction in omnichannel retailing.” *Journal of Retailing and Consumer Services* 61: 102550.

Verma, Prashant, and Shruti Jain. 2015. “Skills augmenting online shopping behavior: A study of need for cognition positive segment.” *Business Perspectives and Research* 3 (2): 126–145.

Vignali, Gianpaolo, Louise F. Reid, Daniella Ryding, and Claudia E. Henninger. 2019. *Technology-Driven Sustainability: Innovation in the Fashion Supply Chain.* Cham: Switzerland: Springer.

Vodolazov, Vlad. 2021. “CLO-Z: The first real-time digital outfit try-on for mobile.” *ARVR Weekly Journal.* Accessed July 19, 2022. https://arvrjourney.com/clo-z-the-first-real-time-digital-outfit-try-on-for-mobile-f5841cc2924b.

Williams, Robert. 2019. “Retailers to spend $1.5B on AR/VR in 2020.” *Retail Dive*, July 19, 2022. www.retaildive.com/news/retailers-to-spend-15b-on-arvr-in-2020/568346/.

Part D

Digital Business and Promotion

8 Direct-to-Consumer Fashion Brands
The Digitally Native Start-Ups in the Retail Industry

Annie Sautel and Sanjukta Pookulangara

CONTENTS

8.1 INTRODUCTION: DIRECT-TO-CONSUMER FASHION BRANDS

The accelerated adoption of digital technologies along with the COVID-19 pandemic have caused the retail landscape to undergo an incredible shift over the past decade. The emergence of advanced technology within the Digital Age has impacted how value is created and perceived by consumers and brands in the retail industry (Gielens and Steenkamp 2019). Due to this digital disruption, the retail industry was on its way to complete digital integration; however, the COVID-19 pandemic accelerated this transition from brick and mortar to digital shopping by almost five years (Ayling 2020). As a result, brands were forced to revaluate their strategies in accordance with consumers' growing preferences for highly immersive and interactive shopping experiences.

DOI: 10.1201/9781003264958-12

Despite traditional retailers adjusting to multichannel formats and adopting digital strategies to meet consumers' changing expectations, a new generation of brands emerged and quickly accomplished this with an innovative and irreplicable business model (Jin and Shin 2020). Direct-to-consumer (DTC) brands are defined as digitally native start-ups who sell directly to consumers without the use of middlemen (Kim et al. 2021). By eliminating costly intermediaries in their value chain, DTC brands can directly interact with consumers, resulting in an improved and highly personalized customer experience (Camarão 2021; Gielens and Steenkamp 2019). Even though DTC brands have been around for nearly ten years, the recent global pandemic surged the relevancy of these brands since consumers shifted their focus to online shopping (Kim et al. 2021). Aside from the growing trend around the globe, the United States has seen a 24.3% increase of DTC sales in 2020 (Botting 2020). Diffusion's (2020) DTC Purchase Intent Index indicates growth with nearly 79% of consumers stating that they plan to purchase more from DTC brands beyond the pandemic. Due to the accelerated emergence of DTC brands, there has been a growing interest in the DTC domain. Therefore, further investigation is needed in order to aid both industry and academic professionals.

8.2 SIGNIFICANCE AND PURPOSE OF STUDY

Preliminary research addressing DTC brands in the retail industry is sparse and limited. This is especially true in terms of research that studies the impact of the COVID-19 pandemic on consumer behaviour and DTC brands. The purpose of this study is to determine the values that influence consumers' attitudes and re-purchase intentions towards DTC brands after the COVID-19 pandemic. The consumption value theory (CVT) will be used as the theoretical framework for this study in order to understand why consumers choose to re-purchase from DTC brands. This theory was modified for the purpose of this study with different variables which include functional value, social value, and epistemic value. These values along with each of their sub-set variables will be tested to measure their impact on attitude which should result in consumers having re-purchase intentions towards DTC brands.

Considering the purpose of this study, there are various implications that are pertinent to both academic and industry professionals. To start, this study will add to the limited amount of research concerning DTC brands in the retail industry. Especially concerning consumer behaviour, research is extremely sparse, so this study will help fill in this gap of literature. This study uses a well-established theory to provide a systematic understanding of consumers' re-purchase intentions for DTC brands. We extracted various variables from the consumption value theory (CVT) for this study's conceptual framework. No previous study regarding DTC brands in the retail industry has utilized this theory, so this study will add a new perspective to the existing literature by incorporating this theory into its research model and discussion. Concerning the current global climate, this study also provides timely implications for brands in the retail industry by providing information regarding consumers' re-purchase intentions towards DTC brands after the global pandemic.

8.3 LITERATURE REVIEW

8.3.1 Emergence of DTC Brands in the Retail Industry

Along with the digitization of the retail industry and COVID-19 pandemic, the rise of DTC brands over the past decade was in response to changing consumer behaviour. The inception of direct-to-consumer (DTC) brands began back in the early 2010s as value-based consumerism drove shoppers to demand high-quality, convenience, affordability, and personalization from brands (Axelsen et al. 2021; The Ultimate Guide 2021). Other factors such as the rise of the internet, growth of online marketplaces, and the demographic shift from baby boomers to millennials also attributed to the emergence of these brands (Corsten et al. 2021; Gielens and Steenkamp 2019; Salpini 2020a). The goal of these digitally native start-ups was to capitalize off under-valued social media platforms and direct distribution strategies that existing incumbents were failing to take advantage of (Schlesinger et al. 2020). With founding dates ranging between 2010–2015, Warby Parker, Casper, Glossier, Away, and Dollar Shave Club are DTC brands that managed to penetrate and seize generous portions of their respective market shares (McGrath 2020). Just like these examples, early DTC entrants focused on offerings that were sold within mature markets such as eyeglasses, shaving, cosmetics, and mattresses (Corsten et al. 2021). For example, when Schnick and Gillette owned 90% of the U.S. shaving market, many consumers shared a distain for the high cost of razors (Pandey 2019; Tiffany 2018). Since then, Dollar Shave Club disrupted the $3.5 billion shaving industry with its DTC subscription service that immediately stole nearly 9% of the market share with its high-quality blades that cost only $1 a month (Tiffany 2018). Just like Dollar Shave Club, early DTC brands that found immediate success are still experiencing strong return today. Companies who wanted to replicate this success have caused the number of DTC brands to accelerate since their emergence (Salpini 2020b). For example, Casper was one of the first DTC brands to sell mattresses in 2014; however, the brand now has more than 200 DTC rivals (Corsten et al. 2021). There was a high growth rate of DTC brands between 2010–2015; however, a slight decline followed as the market began to be saturated by many more DTC competitors (Salpini 2020a; Schlesinger et al. 2020). In 2020, a resurgence occurred as DTC sales grew by 24.3% in response to the COVID-19 pandemic (Botting 2020). This pandemic surged the relevancy of DTC brands because the brick-and-mortar shutdown shifted consumers' focus to online shopping (Kim et al. 2021). Coinciding with the results of Diffusion's (2020) survey, DTC brands will not become obsolete after physical channels begin to re-open because 79% of consumers stated that they intend to purchase from DTC brands even after the pandemic is over. Considering the evolution of DTC brands, there is no dispute that these digitally native start-ups have made a notable impact on brands and consumers within the retail industry.

8.3.2 Consumer Purchase Behaviour in Relation to the DTC Brands

DTC brands successfully emerged in the retail industry due to their ability to offer consumers new value propositions that traditional retailers could not provide (Jin and Shin 2020). DTC brands provide consumers with a valuable consumer experience with their use of highly personalized touchpoints, greater product utility, and reduced customer service friction. After examining preliminary research within the DTC domain, we found that consumer value was highly associated with DTC brands' use of data acquisition (Pichoff 2019). These brands can collect proprietary data directly from their customers which would otherwise be intercepted by intermediaries or third parties (Gielens and Steenkamp 2019). As a result, these brands can leverage consumer data in order to provide a highly personalized experience (Ament 2020). Data such as previous searches, purchases, and browsing history are used by DTC brands in order to gain insight about consumer behaviour (Pichoff 2019). This data can be used for better tailored product recommendations, targeted advertisements, and website navigation (Weinsten 2021).

Many consumers gravitate towards DTC brands because they tend to have a quicker rollout of in-demand products when compared to traditional players (Ament 2020). DTC brands' direct connection to consumer data makes this possible because it gives them the ability to quickly access and respond to customer feedback regarding their needs and concerns (Gielens and Steenkamp 2019). For example, Harry's direct access to their customers' feedback allowed them to make tweaks to their new razors right before the launch date, which would have been otherwise unrealistic if middlemen were in the way (Arora et al. 2020). Glossier relied on customer feedback left on their Instagram posts before developing their next consumer value-driven beauty line, which led to the success of their million-dollar business. Since these brands avoid costs associated with intermediaries, they are also able to offer high-quality products at lower prices when compared to their traditional competitors (Heyward 2021; Kim et al. 2021). Because of this, consumers experience greater product utility which is the perceived value of a product relative to its price and alternative offerings (Maverick 2020).

Consumers also experience greater service utility because DTC brands directly manage customer service interactions (Ament 2020). This will reduce friction and lead to a higher quality service because customers do not have to go through various intermediaries when attempting to resolve their customer service issues (Ament 2020). The direct model puts customers at the centre of the brands' story which strengthens the relationship between the customer and brand and, in turn, maximizes consumer value (Ament 2020; Salpini 2020b). Overall, the value offered by the DTC model allows retailers to strengthen their relationship with consumers by offering a value-driven experience that traditional brands struggle to compete with.

8.4 PREVIOUS RESEARCH IN THE AREA OF DTC

The domain of research regarding DTC brands has undergone both quantitative and qualitative types of investigations. Much of the available research consists of qualitative studies that have collected data through a limited range of methods. For example, the semi-structured interview approach has been a popular tool used by researchers to gather data from DTC brand managers and consumers who are frequent DTC shoppers (Camarão 2021; Kim et al. 2021). Preliminary research on DTC brands in the retail industry has mainly provided a conceptual understanding of the

DTC business model and the various strategies, benefits, and challenges associated with it (Kim et al. 2021). Other research on DTC brands examines specific markets within the retail industry, like DTC brands within the athletic wear or modern-luxury market (Camarão 2021). Additionally, much of the available research offers a conceptual overview of DTC brands from the retailer's perspective (Camarão 2021; Gielens and Steenkamp 2019; Jin and Shin 2020). No study before has investigated DTC brands and consumer behaviour through the lens of the consumption value theory. This theory is used to explore consumer choice behaviour (Sheth et al. 1991). This theory is appropriate to use as the basis for this study's conceptual framework because repurchase intention is a form of consumer consumption choice-oriented behaviour (Kaur et al. 2018). Therefore, the values adopted from this theory along with additional constructs will be used for this study's research model.

8.5 CONCEPTUAL FRAMEWORK

When studying consumer behaviour, consumption values are important to consider because they help explain why consumers make certain choices in various shopping contexts (Sheth et al. 1991). The consumption value theory (CVT) was created in order to identify the values that influence consumer choice behaviour (Sheth et al. 1991). The five consumption values proposed in this theory include functional value, conditional value, social value, emotional value, and epistemic value. In the present research, we have adopted the theoretical framework of the CVT theory by incorporating three out of the five original consumption values in this study's research model. Considering the DTC context, we presume three values as antecedents of consumers' attitudes and repurchase intentions towards DTC brands: functional value, social value, and epistemic value. Additionally, we have adopted a similar approach by Kaur et al. (2018), which considers various constructs that support the chosen values. As seen in Figure 8.1, these constructs are listed under each of its

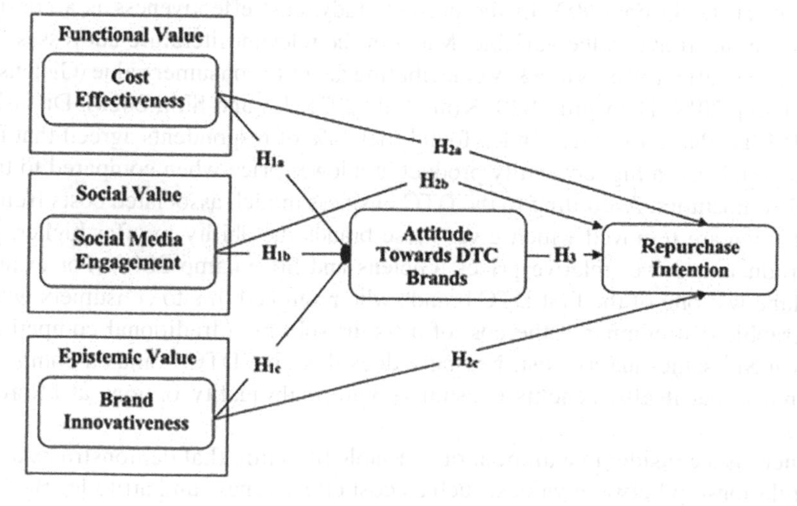

FIGURE 8.1 Conceptual framework and hypotheses.

adjoining value, which include cost effectiveness, social media engagement, and brand innovativeness.

8.6 HYPOTHESIS DEVELOPMENT

Building on the adapted conceptual framework, we propose seven different hypotheses to explain consumers' repurchase intentions towards DTC brands. Functional value is considered as the perceived utility consumers experience from choosing an option that possesses attractive attributes such as reliability, longevity, or fair price (Sheth et al. 1991). Given the prior research about the value of DTC brands, functional value is an appropriate variable for this study's model because the value derived from DTC brands is greatly driven by these brands' functional attributes such as their ability to offer high-quality products at competitive prices (Heyward 2020; Kim et al. 2021). Social value is considered as the perceived utility obtained from a choice that is associated with social groups (Sheth et al. 1991). Social value was also chosen as one of the values because previous research found that social media engagement associated with DTC brands positively influenced consumers' behaviour and attitudes (Kim et al. 2021). Epistemic value is considered as the perceived utility acquired from a choice that arouses curiosity and the need for more knowledge from consumers (Sheth et al. 1991). This was the final chosen value that was appropriate for this study because much of the previous research has consistent findings that note the novelty nature of DTC brands' ability to offer consumers with new and unique solutions (Ament 2020; Kim et al. 2021; Gielens and Steenkamp 2019; Jin and Shin 2020). Under each of these chosen values are the constructs that specifically support each value and how it positively influences consumers' attitudes and repurchase intentions towards DTC brands. Next, we will define the chosen constructs as well as justify the nine proposed research hypotheses.

Cost effectiveness means that the offerings have good value for the given price (Oliver and DeSarbo 1988). In the present study, cost effectiveness is a construct under the functional value variable. Much of the relevant literature addresses DTC brands' cost-effectiveness as a key contributing factor to consumer value (Gielens and Steenkamp 2019; Heyward 2020; Kim et al., 2021; Jin and Shin 2020). Diffusion's (2021) DTC Purchase Intent Index found that 44% of respondents agreed that DTC brands produced a higher quality product at a lower price when compared to traditional competitors. According to the DTC business model, associated costs from the middleman are removed which gives these brands the ability to offer higher quality products at lower relative prices (Gielens and Steenkamp 2019). For example, Everlane was one of the first DTC brands who promoted this to consumers with an infographic that compared the cost of a t-shirt sold by a traditional competitor to its own $15 same-quality cost. Not only does this give DTC brands a competitive advantage, but it also benefits consumers with high-quality options at affordable prices.

There is a considerable amount of available literature that demonstrates a positive relationship between values, such as cost effectiveness and attitude (Baek and Oh 2020). In the DTC domain, various researchers have findings that support this relationship. For example, Kim et al. (2021) found that cost effectiveness had a positive effect on consumer attitude towards DTC brands. Not only did they find a positive relationship, but cost effectiveness was shown to have the strongest impact on attitude out of all the determinants studied (Kim et al. 2021). Going further, Ajzen and Fishbein's (1980) Theory of Reasoned Action states that attitude is one of the key determinants of consumer behaviour. Values that impact attitudes have been investigated by a considerable number of researchers to predict consumer behaviour (Sheth et al. 1991). Based on these findings, we propose the following hypotheses:

H1a. Cost effectiveness of DTC brands will positively influence attitude towards DTC brands.

H2a. Cost effectiveness of DTC brands will positively influence repurchase intentions of DTC brands.

Social media engagement is described as consumers' behavioural manifestations on social media platforms. According to Hoffman and Fodor (2010), consumer brand engagement is a customer's level of investment in a brand based on cognitive, emotional, and behavioural levels. In other words, consumers feel satisfied, loyal, emotionally attached, and connected to brands that they are engaged with on social media. In the DTC context, social media is a crucial tool for these brands to use when communicating with customers about informational and promotional messages (Schlesinger et al. 2020). Additionally, DTC brands are known for having online brand communities that consist of loyal consumers who engage with one another over various social media platforms. In the context of consumer engagement with the Glossier DTC brand, Paintsil (2019) found that consumers gained value from the two-way conversations facilitated by the brand's online social media community.

Previous research on topics related to social media engagement has received a considerable amount of attention in the academic community (Kaur et al. 2018; Schivinski and Dabrowski 2016; Schlesinger et al. 2020). A limited amount of this literature directly examines DTC brands; however, the findings are applicable because the DTC business model is heavily invested in various forms of social media engagement. Previous studies have identified a positive link between social media engagement, consumer attitude, and repurchase intentions (Kim et al. 2021; Schivinski and Dabrowski 2016). Therefore, we propose the following hypotheses:

H1b. Social media engagement of DTC brands will positively influence attitude towards DTC brands.

H2b. Social media engagement of DTC brands will positively influence repurchase intentions of DTC brands.

Brand innovativeness is perceived by consumers as the extent to which brands can fulfil their needs with new and useful solutions (Alpert et al. 2015). In the present study, brand innovativeness is a construct under the epistemic value variable.

Consumers can obtain epistemic value in different ways, such as gaining information or by experiencing something new (Sheth et al. 1991). Previous research has consistent findings in line with consumers experiencing value from DTC brands' ability to provide innovative solutions (Gielens and Steenkamp 2019; Kim et al. 2021; Jin and Shin 2020; Paintsil 2019). Without the middleman, DTC brands have direct access to consumer data. This proprietary information is leveraged by these brands to drive innovation with solutions and unique products that are not offered by competitors (Gielens and Steenkamp 2019). For example, a major hurdle in the eyewear market was the inability for customers to try on glasses before purchasing online. Warby Parker addressed this problem with its "Home Try-On" service which allowed consumers to try on products without incurring additional costs such as return shipping fees (Jin and Shin 2020; Ament 2020). This is just one example of many innovative solutions DTC brands have come up with over the years to add epistemic value to their offerings.

Even though research regarding consumer behaviour and DTC brands is sparse, much of the literature acknowledges the value consumers derive from DTC brands' innovativeness (Axelsen et al. 2021; Arora et al. 2020; Gielens and Steenkamp 2019; Kim et al. 2021; Jin and Shin 2020). For the purpose of this study, brand innovativeness was adopted as a construct from Kim et al.'s (2021) research, which found a positive link between perceived brand innovativeness and consumer attitude towards DTC brands. As previously stated, attitude is one of the key determinants of consumer behaviour (Ajzen and Fishbein 1980). The positive link between attitude and behavioural intentions has been evaluated in the DTC research domain. Arora et al. (2020) reported that nearly two-thirds of consumers had intentions to repurchase from DTC brands after the COVID-19 pandemic was over. Additionally, Diffusion's 2021 DTC Purchase Intent Index report found that 79% of consumers who were already familiar with DTC brands had plans to continue to purchase from them after 2021 (DeLaite 2021). Going further, Kim et al. (2021) also found a positive correlation between DTC brand innovativeness and consumers' repurchase intentions. Considering these arguments, we propose the following hypotheses:

H1c. Brand innovativeness of DTC brands will positively influence attitude towards DTC brands.

H2c. Brand innovativeness of DTC brands will positively influence repurchase intentions of DTC brands.

A previous study conducted by Kim et al. (2021) investigated consumers' behaviour and repurchase intentions towards DTC brands. This study, along with others in the same domain, found a positive link between consumers' attitudes and willingness to purchase from DTC brands again in the future (Camarão 2021; Kim et al. 2021). Therefore, we propose in our final hypotheses that the values, constructs, and attitude will positively influence repurchase intentions of DTC brands (Table 8.1).

H3. Attitude towards DTC brands will positively influence repurchase intentions of DTC brands.

TABLE 8.1
Summary of Definition and Relevant Literature for Each Construct

Values and Constructs	Definition	Relevant Literature
Cost Effectiveness	Offerings are "good value for the money" (Oliver and DeSarbo 1988).	DTC brands' offerings have high perceived value or "consumer's overall assessment of the utility of product (or service) based on perceptions of what is received and what is given", relative to alternate offerings (Zeithaml 1988, 14; Kim et al. 2021).
Brand Innovativeness	The extent to which consumers perceive brands as being able to provide new and useful solutions to their needs. The innovation can include business models, products, storytelling, and all other brand activities (Eisingerich and Rubera 2010, 66; Schlesinger et al. 2020).	Direct interaction with consumers allows DTC brands to gather proprietary information (otherwise mediated by retailers/middleman) which drives innovation (Gielens and Steenkamp 2019).
Social Media Engagement	Social media is often the primary channel for marketing; hyperactive brand–customer interaction through social media; customers' behavioural manifestations in social media beyond purchase as an online brand community (Kim et al. 2021).	Social media is a critical communication tool for DTC brands to interact with their customers, conveying both promotional and informational messages, and offering a platform for their customers to communicate (Schlesinger et al. 2020).
Repurchase Intention	Consumers' intention to repurchase a product or continue service use is determined primarily by their satisfaction with prior use of that product or service (Bhattacherjee 2001).	79% of those familiar with DTC brands say they plan to increase their DTC purchases in 2021 (DeLaite 2021).
Attitude	Theory of Planned Behaviour—the attitude towards a behaviour influences and behavioural intention (Ajzen 1991).	52% of consumers say that 20% or more of their 2021 purchases will be from DTC brands (DeLaite 2021).

8.7 DATA COLLECTION

After determining the constructs and hypothesized relationships between the variables, a survey was conducted to test its validity. The online survey was designed and administered from the web-based software Qualtrics. The software established a link that was sent to those over email and through social media to complete the survey on the Qualtrics platform. Two micro-influencers were chosen to post the link to the survey due to their previous sponsorships and engagement with DTC brands. Emails containing a description of the research and a request for them to participate were sent to these influencers. After getting their confirmation and receiving approval from the Institutional Review Board (IRB), a self-administered, only survey link was established and sent to willing participants. The survey was active for multiple weeks and counted 444 respondents when the data was ready to be extracted. For screening purposes, two questions asked respondents if they have purchased from DTC brands before and whether they have purchased from the proposed list of 12 popular DTC brands. From those initial questions, only those who acknowledged purchasing from DTC brands were able to complete the survey. At the end of the survey, there were questions regarding the respondents' demographic information which is provided in Table 8.2. Data was later collected after a three-week time frame and resulted in a random sample of n = 204.

8.8 INSTRUMENT DEVELOPMENT

The items and scales were adapted from previous studies to measure the proposed hypotheses of consumers' attitude and repurchase intentions towards DTC brands. The survey contained five variables: cost effectiveness, social media engagement, brand innovativeness, attitude towards DTC brands, and repurchase intention towards DTC brands. The variables were measured using multi-scale items that were also adapted from previous studies. Items were measured using the five-point Likert

TABLE 8.2
Demographic Characteristics of the Sample

Variable	N = 204	Frequency
Gender		
Male	54	26.5%
Female	145	71.1%
Non-binary/third gender	3	1.5%
Prefer not to say	2	1%
Age		
18–25	83	40.7%
26–35	35	17.2%
36–49	52	25.5%
50–65	30	14.7%
65+	4	2%
Education		
Less than high school	1	.5%
High school graduate	16	7,8%
Some college	24	11.8%
2-year degree	9	4.4%
4-year degree	91	44.6%
Professional degree	53	26%
Doctorate	8	3.9%
Prefer not to say	2	1%

scale ranging from "strongly disagree" to "strongly agree". A factor analysis was conducted using SPSS to test the variability of the survey questions. All factor items were retained if the loadings were greater than .50. Only considering factor loadings greater than .50, all items were retained. All five components were extracted with Eigenvalues greater than one. Out of these five components, the variable with the strongest factor loading was attitude followed by repurchased intention, social media engagement, cost effectiveness, and ending with brand innovativeness.

8.9 RESULTS

After conducting the factor analysis, multiple regression was performed to understand how the variables within the purposed hypotheses impact one another. The results of the regression analysis are summarized in Table 8.3 and Figure 8.2.

The results show that cost effectiveness, social media engagement, and brand innovativeness all positively influence attitude towards DTC brands, supporting H1a, H1b, and H1c. Cost effectiveness, social media engagement, and brand innovativeness can predict 35.2% of the variance in attitude towards DTC brands. The most powerful predictor of attitude towards DTC brands was cost effectiveness ($\beta = .339$, $p =. 000$), followed by brand innovativeness ($\beta = .273$, $p = .000$), and then social media engagement ($\beta = .211$, $p = .001$). Similarly, all three variables positively influenced repurchase intentions of DTC brands, supporting H2a, H2b, and H2c. Cost effectiveness, social media engagement, and brand innovativeness can predict 43.6% of the variance in repurchase intention. The most powerful predictor of repurchase intentions of DTC brands was brand innovativeness ($\beta = .373$, $p = .000$), followed by cost effectiveness ($\beta = .317$, $p = .000$), and then social media engagement ($\beta = .210$, $p = .000$).

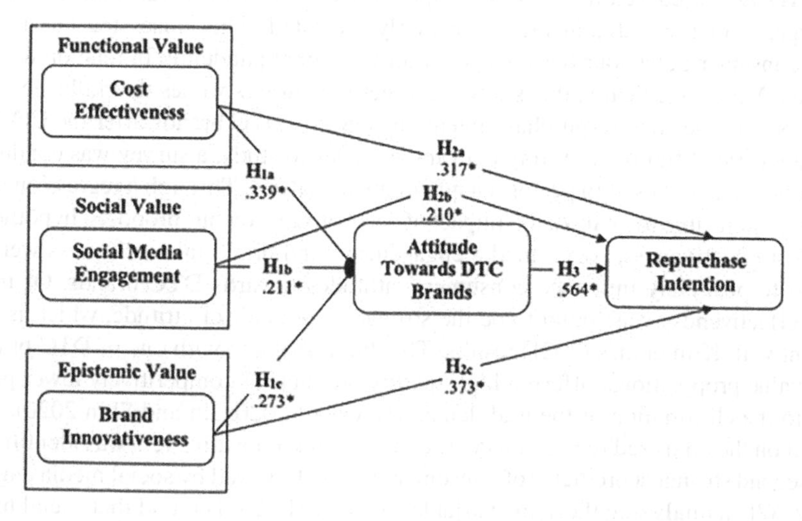

FIGURE 8.2 Results of the conceptual model.

Note: The beta coefficients for each variable are listed by the corresponding hypotheses. The asterisk indicates $p \leq .001$.

TABLE 8.3
Regression Results between the Studied Variables

Factor Items	Adjusted R Squared	Coefficients Beta	Significance
Attitude	.352		
Brand Innovativeness		.273	.000
Cost Effectiveness		.339	.000
Social Media Engagement		.211	.001
Repurchase Intention	.436		
Brand Innovativeness		.373	.000
Cost Effectiveness		.317	.000
Social Media Engagement		.210	.000
Repurchase Intention	.315		
Attitude		.564	.000

Finally, H3 was also supported because attitude towards DTC brands was found to be a strong predictor ($\beta = .564$, $p = .000$) of repurchase intentions of DTC brands. Attitude can predict 31.5% of the variance in repurchase intentions of DTC brands.

8.10 DISCUSSION

The retail industry is constantly evolving in accordance with consumers' changing needs and preferences. As a result, direct-to-consumer brands have emerged to meet these demands with unique offerings that traditional retailers are unable to provide. COVID-19 has caused an incredible ripple effect within the retail industry, so the emergence of these digital brands has only escalated. This raises the question of how consumer behaviour has changed after the global pandemic in relation to DTC brands. More specifically, this study examines the various values that influence consumers' attitudes and re-purchase intentions towards DTC brands after the COVID-19 pandemic. After determining relevant variables to study, a survey was conducted to gather data and used to perform a quantitative analysis. Through a regression analysis, we were able to analyze the impact of each variable on our proposed hypotheses.

Cost effectiveness, social media engagement, and brand innovativeness were all found to positively influence consumers' attitudes towards DTC brands. Of these, cost effectiveness was found to be the strongest predictor of attitude, which is consistent with Kim et al.'s (2021) results. This finding aligns with one of DTC brands' key value propositions: offering high-quality products at competitively lower prices due to the elimination of the middleman (Heyward 2020; Jin and Shin 2020). Also based on the analyzed regression coefficients, brand innovativeness was identified as the second strongest predictor of consumer attitude followed by social media engagement. When analyzing the same variables, Kim et al. (2021) found that social media engagement was the second strongest predictor, and brand innovativeness was one of the weakest predictors of attitude. Even though both studies agree that the variables positively influence attitude, this discrepancy means that the two samples in the different studies are likely to be influenced to a greater degree depending on the variable. This finding suggests that consumer behaviour has shifted after the global pandemic to value innovativeness higher than the social media engagement associated with DTC brands. This aligns with recent post-pandemic studies that demonstrate findings regarding consumer preferences shifting in accordance with DTC brands' ability to offer innovative solutions (DeLaite 2021).

Going further, all four of the independent variables in H2a, H2b, H2c, and H3 were found to positively influence consumer repurchase intentions of DTC brands. To start, there was a positive correlation between consumer attitude and repurchase intention which is in line with prior research. Out of cost effectiveness, social media engagement, and brand innovativeness, the strongest predictor of repurchase intention was brand innovativeness. This finding suggests that DTC brands' ability to fulfil consumers' needs with innovative solutions motivates consumers to purchase their products again in the future. Following brand innovativeness, cost effectiveness was the second strongest predictor followed by social media engagement. This finding is interesting because compared to the other two factors, social media engagement was rated as the least strong predictor in both attitude and repurchase intention. This indicates that consumers are more likely to have a positive perception towards DTC brands due to their innovative offerings that have high perceived value for reasonable prices (Heyward 2020; Kim et al. 2021). Above all else, these results confirm that consumers prefer brands that are cost-effective and can meet their ever-changing needs and demands.

8.11 THEORETICAL AND MANAGERIAL IMPLICATIONS

There is a limited amount of research in academia regarding DTC brands in the retail industry. Since these brands have only relatively recently emerged in the retail industry, much of the available literature only provides a conceptual overview of the various DTC models. Additionally, other types of existing research have only scratched the surface when it comes to analyzing consumer behaviour in relation to DTC brands. Especially looking through the lens of the CVT theory, this research is one of the first of its kind to study consumer behaviour before and after the pandemic. Due to COVID-19, many different aspects of consumer behaviour have been impacted. This study is one of the first of its kind to acknowledge this shift and study its impact on consumer attitudes and re-purchase intentions towards DTC brands.

This study is especially pertinent to professionals within the retail industry. DTC brands can take advantage of the findings when marketing their value propositions to consumers. For example, since brand innovativeness was found to be the strongest predictor of repurchase intention, then a DTC brand could market its need-fulfilling benefits to consumers. As a result, consumers will be more likely to purchase more from the brand in the future. Non-DTC retailers can also benefit from this research. Not only does this study explain the factors that draw consumers to these brands, but it also provides necessary background information about DTC brands and their relevancy in the retail industry pre and post COVID-19. This can help traditional retailers make strategic decisions if they are considering entering the DTC market. Additionally, it can help these retailers understand DTC brands' strategies so that they can compete with them better.

8.12 LIMITATIONS

There are limitations to this study, which suggests various directions for future research. To start, the sample used for this study was general in terms of its demographic makeup. Data was gathered from participants who represented a broad range of ages, level of education, and various backgrounds. Since this study was not aimed towards a specific demographic, future studies can look more closely at certain consumer groups, like the values of Generation Z or millennials. Additionally, a future study can analyze different geographic regions. Since this study was geared towards American consumers, future research can gather data from other groups outside of the United States. This study focused on consumer behaviour, and more specifically, it analyzed the values impacting consumer attitudes and repurchase intentions. Future studies could analyze different aspects of consumer behaviour like the various deterrents that might cause consumers to avoid DTC brands. Finally, future studies could examine the impact COVID-19 had on DTC brands through the perspective of the retailer instead of the consumer. All retailers had to adjust their strategies this past year, so investing this shift from the brand's perspective could contribute valuable insight to academic and industry professionals.

REFERENCES

Ajzen, I. 1991. "The theory of planned behavior." *Organizational behavior and human decision processes*, 50(2), 179–211.

Ajzen, I., and M. Fishbein. 1980. *Understanding Attitudes and Predicting Social Behavior.* Englewood Cliffs: Prentice-Hall. Inc,.

Alpert, F., M. Brown, and R. Shams. 2015. "Consumer perceived brand innovativeness." *European Journal of Marketing* 49 (9/10): 1589–1615.

Ament, L. 2020. "Is your enterprise ready for direct-to-consumer business?" *Customer Relationship Management* 24 (2): 9. www.destinationcrm.com/Articles/ReadArticle.aspx?ArticleID=139447.

Arora, A., H. Khan, and S. Kohli. 2020. "DTC e-commerce: How consumer brands can get it right." *McKinsey & Company.* Accessed October 24, 2022. www.mckinsey.com/capabilities/growth-marketing-and-sales/our-insights/dtc-e-commerce-how-consumer-brands-can-get-it-right.

Axelsen, E., J. Kim, B. Christiansen, L. Yang, A. Zhong, J. Doolan, A. Mao, and S. Mohammadzadeh. 2021. "Fashion: COVID-19 and social trends." *Harvard College Consulting Group.* Accessed October 24, 2022. https://static1.squarespace.com/static/5edeedaed417cb4da8b5864a/t/600fabdc5c129e5b3cbad944/1611639772893/%5BHCCG%5D+Inclusive+Fashion+Report.pdf.

Ayling, A. 2020. "How has the pandemic impacted digital retail?" *Insights.* Accessed October 24, 2022. https://insights.bu.edu/how-has-the-pandemic-impacted-digital-retail/.

Baek, E., and G. Oh. 2020. "Diverse values of fashion rental service and contamination concern of consumers." *Journal of Business Research* 123 (2021): 165–175. https://doi.org/10.1016/j.jbusres.2020.09.061.

Bhattacherjee, A. 2001. "Understanding information systems continuance: An expectation-confirmation model." *MIS quarterly*, 25(3), 351–370.

Botting, N. 2020. "Direct-to-consumer booms during covid-19." *IMRG: The UK Ecommerce Association.* Accessed October 24, 2022. www.imrg.org/blog/direct-to-consumer-booms-during-covid-19/.

Camarão, R. 2021. *The Evolution of the Direct-to Consumer Model in the Sportswear Industry: An Assessment of the Model's Application by Top Industry Leaders* (Doctoral dissertation). The Catholic University of Portugal. Google Scholar. https://repositorio.ucp.pt/bitstream/10400.14/34966/1/152119204_Rafael%20Camarao_DPDFA.pdf.

Corsten, D., M. Higgins, K. Rangan, and L.A. Schlesinger. 2021. "How direct-to-consumer brands can continue to grow." *Harvard Business Review.* Accessed October 24, 2022. https://hbr.org/2021/11/how-direct-to-consumer-brands-can-continue-to-grow.

DeLaite, B. 2021. "Traditional brands shift to DTC to combat lack of retail traffic." *Forbes.* Accessed October 04 February 2023, https://www.forbes.com/sites/forbescommunicationscouncil/2021/03/05/traditional-brands-shift-to-dtc-to-combat-lack-of-retail-traffic/?sh=6a86ed846a38.

Diffusion. 2020. "Diffusion's 2021 direct-to-consumer purchase intent index: DTC is going mainstream." Accessed October 24, 2022. https://diffusionpr.com/us/2021-dtc-purchase-intent-index/.

Eisingerich, A. B., & Rubera, G. 2010. "Drivers of brand commitment: A cross-national investigation." *Journal of International Marketing*, 18(2), 64–79.

Gielens, K., and Steenkamp, J. 2019. "Branding in the era of digital (dis)intermediation." *International Journal of Research in Marketing* 36 (3): 367–384. https://doi.org/10.1016/j.ijresmar.2019.01.005.

Heyward, E. 2020. "How the smartest DTC brands will evolve, and what they can teach you." *Inc.* Accessed October 2022. www.inc.com/magazine/202006/emily-heyward/direct-to-consumer-dtc-digital-native-brand-ecommerce-red-antler.html.

Hoffman, D. L., & Fodor, M. 2010. "Can you measure the ROI of your social media marketing?" *MIT Sloan management review*, 52(1), 41–49.

Jin, B.E., and D.C. Shin. 2020. "Changing the game to compete: Innovations in the fashion retail industry from the disruptive business model." *Business Horizons* 63 (3): 301–311. https://doi.org/10.1016/j.bushor.2020.01.004.

Kaur, P., A. Shir, R. Rajala, and Y. Dwivedi. 2018. "Why people use online social media brand communities a consumption value theory perspective." *Online Information Review* 42 (2): 205–221.

Kim, N.L., D.C. Shin, and K. Gwia. 2021. "Determinants of consumer attitudes and re-purchase intentions toward direct-to-consumer (DTC) brands." *Fashion and Textiles* 8 (1): 1–22. http://dx.doi.org.libproxy.library.unt.edu/10.1186/s40691-020-00224-7.

Maverick, J.B. 2020. "What are the four types of economic utility?" *Investopedia.* Accessed October 24, 2022. www.investopedia.com/ask/answers/032615/what-are-four-types-economic-utility.asp.

McGrath, R.G. 2020. "The new disrupters." *MIT Sloan Management Review* 61 (3): 28–33. https://sloanreview.mit.edu/article/the-new-disrupters/.

Oliver, R. L., & DeSarbo, W. S. 1988. "Response determinants in satisfaction judgments." *Journal of consumer research*, 14(4), 495–507.

Paintsil, A. 2019. "*Consumer engagement with modern luxury direct-to-consumer brands on social media: A study of Glossier.*" Master's Thesis, University of Delaware. Retrieved from https://udspace.udel.edu/items/b72d963e-33d4-42f4-8631-c14138def46a.

Pandey, E. (2019, May 11). Shaving giants sweep up the disrupters. *Axios.* https://www.axios.com/2019/05/11/shaving-giants-schick-gillette-harrys-dollar-shave-club-acquisition

Pichoff, S. 2019. "How the direct-to-consumer model is driving retail supply chain innovation." *Retail Info System.* Accessed 24, 2022. https://risnews.com/how-direct-consumer-model-driving-retail-supply-chain-innovation.

Salpini, C. 2020a. "Is the DTC brand aesthetic bad for business?" *Retail Dive.* Accessed 24, 2022. www.retaildive.com/news/is-the-dtc-brand-aesthetic-bad-for-business/588062/.

Salpini, C. 2020b. "The decade of VC funding that shaped e-commerce and DTC brands." *Retail Dive*. Accessed 24, 2022. www.retaildive.com/news/the-decade-of-vc-funding-that-shaped-e-commerce-and-dtc-brands/587134/.

Schivinski, B., and D. Dabrowski. 2016. "The effect of social media communication on consumer perceptions of brands." *Journal of Marketing Communications* 22 (2): 189–214.

Schlesinger, L., M. Higgins, and S. Roseman. 2020. "Reinventing the direct-to consumer business model." *Harvard Business Review*. Accessed October 24, 2022. https://hbr.org/2020/03/reinventing-the-direct-to-consumer-business-model.

Sheth, J.N., B.I. Newman, and B.L. Gross. 1991. "Why we buy what we buy: A theory of consumption values." *Journal of Business Research* 22 (2): 159–170.

Tiffany, K. 2018. "The absurd quest to make the 'best' razor." *Vox*. Accessed October 24, 2022. www.vox.com/the-goods/2018/12/11/18134456/best-razor-gillette-harrys-dollar-shave-club.

The ultimate guide to direct to consumer (DTC). Brightpearl. Accessed April 20, 2021. www.brightpearl.com/sales-channel-strategy/direct-to-consumer-dtc.

Weinsten, L. 2021. "What retailers have learned from DTC strategies (and vice versa)." *Gimbal*. Accessed October 24, 2022. https://gimbal.com/retail-dtc-strategies/.

Zeithaml, V.A. 1988. "Consumer perceptions of price, quality, and value: A means-end model and synthesis of evidence." *Journal of Marketing* 52 (3): 2–22.

9 Fashion Marketing with Virtual Humans as Influencers

Evrim Buyukaslan Oosterom, Fatma Baytar and Mona Maher

CONTENTS

9.1 INTRODUCTION

In October 2021, the social networking platform Facebook was renamed as Meta. Mark Zuckerberg, CEO of the company, spoke at the company's annual virtual reality conference and explained the plans to evolve the company into a metaverse venture (Isaac 2021). Since Zuckerberg's speech, the public has eyes flicked over to the theme metaverse. As Bauld (2022) described it, the metaverse is an updated version of the Internet where people can extend beyond their physical selves by using immersive technologies such as augmented reality (AR) and virtual reality (VR). Extending beyond the physical self requires the user to be represented virtually in internet space, which brings us to a further topic: virtual humans. Before dwelling on this phenomenon, it is essential to understand the creation of the digital image, which is merely a product of computer graphics.

Computer-generated imagery (CGI) is the process of creating static or dynamic content using computer software (Nashville Film Institute n.d.). Although CGI has

DOI: 10.1201/9781003264958-13

been used in the game and film industry for many years, it has emerged in the fashion industry only recently. Lil Miquela, one of the first computer-generated humans (CGH) who became an Instagram phenomenon, made her debut in 2016. She modelled for fashion brands, such as Calvin Klein and Prada, and had acquired 3 million followers as of May 2022. In 2018, Lil Miquela, described as a robot on her Instagram page, was among the 25 most influential "people" on the Internet (TIME staff 2017). Since then, other VIs have been introduced. According to VirtualHumans (n.d.), there are 175 acknowledged VIs worldwide currently (35 of which are Instagram verified).

In such industries as fashion, technology, and music, VIs are employed typically to endorse a product, enhance brand recognition, or raise awareness of a social matter (VirtualHumans 2020). For example, Puma created its own virtual human, Maya, in 2020 to promote a shoe drop (Farveen 2020). Later, Prada created its virtual muse, Candy, to advertise its newly launched fragrance of the same name (VirtualHumans 2021a, 2021b). The global influencer marketing volume is estimated to reach $16.4 billion in 2022 with more than a 500% increase in the last five years (Santora 2022). As VI marketing is a recent phenomenon, there are no reliable statistical data to determine its market value yet. However, with the increasing number of VIs and partnerships with them, a market size escalation can be predicted.

In this chapter, VIs, particularly those that are human-like, are explored in detail to better understand their effects on fashion marketing. A comprehensive literature review was conducted to identify the prominent themes associated with VIs. Based upon the literature review, three focal themes in VIs emerged: (1) creation and technology behind virtual humans, (2) the social interaction between VIs and real people, and (3) VIs' product endorsement and brand advertising, and followers' response to it. The following sections combine these themes to provide a broad understanding of the topic.

9.2 WHAT IS A VIRTUAL HUMAN?

When the word "virtual human" is entered into a web search engine, the results reveal a plethora of definitions of the term. Computer-generated humans, virtual characters, virtual agents, avatars, digital humans, and artificial humans are among the definitions that many academic and non-academic sources use interchangeably. Even though there are some differences among them, a systematic naming of these individual entities has not been done so far. In Figure 9.1, these differences are categorized by filtering academic and non-academic usage of these words. CGI is an umbrella term for all these terms and can be described as using 3D computer graphics to create visual content, such as characters, scenes, or special effects, and this static or dynamic visual content can be the subject of a movie, video game, TV programme, or any other digital or printed media (Rehak 2011). A CGH, which is a subset of CGI, is a virtual character that has the appearance of a real human and is created using 3D computer graphics. On the other hand, a virtual character may not have the appearance of a real human. Gollum, in the Lord of the Rings, can be considered a good example of this. Virtual agents are like virtual characters, but they often signal that there is an interaction between the virtual character and the user/follower.

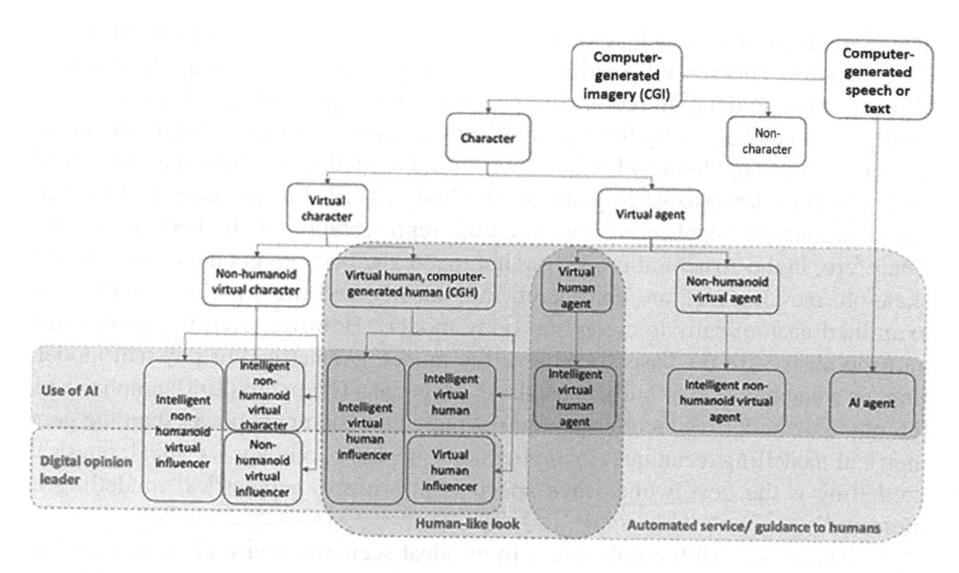

FIGURE 9.1 Categorization of virtual individuals.

Virtual and digital humans may mean the same thing depending on the context and can be used interchangeably with CGH, possibly because they sound less technical and appeal more to the public. Lastly, artificial humans, sometimes referred to as AI humans or intelligent humans, are either created or interact using AI technology. These virtual characters can sometimes be so realistic that they are nearly indistinguishable from real humans (Hill and White 2020). Further to their realistic visual characteristics, they may represent a real human's emotional or behavioural characteristics (e.g., being surprised, sad, or happy) (Li et al. 2021). The customer service field in particular is taking advantage of artificial humans' capabilities because they can interact with customers, answer questions, and help solve their problems; thus, they are referred to sometimes as virtual assistants.

9.3 TECHNOLOGY USED TO GENERATE VIRTUAL HUMANS—3D BODY MODELLING

Two aspects of a virtual human can be used in marketing: cognitive (i.e., communication, perception, language) and physical (i.e., appearance, including the body, face, hair, and dress). Creating a virtual human with a realistic appearance is the subject of collaboration among different research fields, including computer science, physics, mathematics, anatomy, anthropometry, materials science, and apparel design. In the following sections, the technical details of a body model's formation are summarized and followed by an explanation of the commercial software programs available to create virtual humans.

Magnenat-Thalmann (2010) and her research group have made significant contributions to modelling and simulating bodies and garments. They classified the body modelling techniques based upon the approaches used: mathematical and structural.

The mathematical approach generates body models using either geometrical shapes (i.e., cylinders, spheres, ellipsoids) as the starting point and deforming these shapes to achieve the model required or using certain physics principles, such as the mass-spring model and elasticity theory, in which various forces play a critical role in the model's formation. On the other hand, structural modelling focuses on anatomy and anthropometry knowledge to generate the body model. For instance, in biomedical or ergonomics applications, an accurate representation of the body is critical. Therefore, in the structural modelling approach, the body components, such as the skeleton, muscles, etc., are considered individually, and the muscles' postures are examined anatomically to create the body models. However, even if anatomy and anthropometry are the basis to achieve the model, geometrical or physical modelling, or a combination of both, is used. As Magnenat-Thalmann (2010) emphasized, the physical modelling technique requires more computational power than the geometrical modelling technique. In applications in which accuracy is critical, physical modelling is the best, while when speed is prioritized, geometrical modelling is preferred.

As Zhu et al. (2019) emphasized, in an ideal scenario, a virtual human (digital human as referred to in their article) should possess all the features of a real human, including its appearance, action, psychology, and physiology, but this ideal scenario is not currently possible because of the technical limitations. Instead, creators of virtual humans focus on one or two of these aspects depending upon different needs and develop virtual humans accordingly. Further to Magnenat-Thalmann's (2010) technical classification, there are alternative ways to classify digital human modelling in which cognitive aspects also become significant. For instance, Zhu et al. (2019) classified digital human modelling according to the digital humans' end use, that is, cognitive or physical. Cognitive digital humans, which are applied in social sciences, concentrate on the interaction between digital and real humans at the level of emotions, verbal communication, etc. Although they are not real, physical-digital humans (robots) focus on mimicking a human body's static or dynamic anatomy, and hence, are the subject of such fields as biomedicine and automotive safety, among other fields.

In another classification approach, Kim and Chung (2015) chose benefits from the method of creating 3D models: non-template- and template-based. 3D capturing devices (e.g., 3D body scanners) or 3D body modelling software (e.g., Blender or Daz3D) can be used to create non-template-based models (Lee and Song 2021). On the other hand, template-based models adjust the shape of an existing humanoid template to fit the shape of a desired human body best. The body models can be in various forms: 3D, 2D, and parametric. They can be 3D model-based (target body obtained using such 3D capturing devices such as 3D body scanners) (Brownridge and Twigg 2014; Jiang et al. 2014); 2D image-based (target body obtained using 2D images such as photographs of the body from different angles) (Zhu and Kwok 2013; Gu et al. 2019), and parametric-based (target body obtained using such body measurements as hip and waist circumference, etc.) (Baek and Lee 2012). As the name implies, the most realistic body models are achieved with a 3D model, as 3D capturing devices can catch the details of complex geometries.

With respect to commercial 3D modelling software programs, the animation and game sectors have been the driving force for initiating and advancing the computer graphics necessary to create virtual characters since the 1970s (Sturman 1994). Since then, many commercial software programs have been released that allow users to develop imaginative (or realistic) characters for movies or games. By 2022, the most popular software programs used to create virtual humans include Blender, Daz Studio, Unreal Engine, and others. Each of these software programs has its advantages and disadvantages. For instance, even though Blender is an open-source free access programme, it is not specific for modelling body forms, whereas Daz Studio specializes in human body modelling but lacks texturing features (i.e., defining the surface details by dressing a 3D model with 2D images). Unreal Engine's MetaHuman Creator Suit is a free-access cloud-based service which enables non-experienced users to create realistic human models easily, but the human models to be created are limited to the human library offered by MetaHuman Creator Suit.

Further to the 3D body modelling techniques and the popular software programs explained previously, deepfake technology has become the one most familiar to the public with respect to the creation of CGH. In 2019, the Face Swap app became highly popular when users swapped their faces with famous actors, singers, etc. The technology used in the deepfake is machine learning and AI (Westerlund 2019). Many VIs, such as Lil Miquela, are created using deepfake technology, in which her CGI face is attached to a real human's body (VirtualHumans 2021c).

9.4 MARKETING WITH VIRTUAL INFLUENCERS (VIs)

Powers and Greenwell (2017) made a thought-provoking statement in their article that our bodies are now promising avatars for brands. Since modern times, our bodies have reflected our souls, will, and moral identity (Goldstein 1992, 75), and now they function as brands as well. As De Perthuis and Findlay (2019) mentioned, similar to fashion photography in the 1960s, influencers are the objects of interest today. Their bodies, clothes, and the arrangement of scenery give messages to their followers. According to Djafarova and Rushworth (2017), influencers' power is rooted in their authenticity and familiarity with people. Influencers act as a bridge between the brands and the consumers by transferring the brands' messages and increasing consumers' ability to relate to the brand (Uzunoğlu and Kip 2014).

Influencers are the new generation of online opinion leaders (Casaló et al. 2020). Opinion leaders are people whose ideas, insights, and fame are valued by the members of a community and have the power to influence others' attitudes toward novel products (Myers and Robertson 1972). This definition is similar to the definition of "influence". Brown et al. (2015) defined an influencer as a third party who affects consumers' decisions whether to purchase a product and cannot be held accountable for the decisions they make. The opinion leader notion extended to the influencer concept with the introduction of Web 2.0, which enabled social networking among users. Personal blogs were the early forms of social media through which influencers affected followers' decisions significantly (Li and Du 2011), and they became even more important with YouTube and Twitter's establishment in 2005. Finally, social media influencers boomed in 2010 with the debut of Instagram (Burns 2021).

These influencers create content for social media channels (i.e., Instagram, YouTube, TikTok, Twitter) and become figures of interest most often because of their social media appearance. This distinguishes social media influencers from celebrity influencers who acquire fame by acting, singing, performing, etc.

Influencer marketing witnessed accelerated growth after 2020, and influencers got more than the estimated number of followers because the COVID-19 pandemic caused people to spend more time online. Although we do not know yet whether the tendency to spend more time online will be permanent, it is known that the number of social media users is increasing considerably every year; for instance, the number of Instagram users reached 1890 million in 2021 from 1210 million in 2019 (Iqbal 2022). Influencer marketers measure digital influencers' effect on people often according to engagement metrics, such as click-through rates, the number of likes, comments, followers, reshares, etc. (McCann and Barlow 2015). However, from a business perspective, the more important factor is whether the interactions between the followers and influencers turn a profit (Santiago and Castelo 2020). Although the statistics on the increased sales attributable to influencers are scarce, Influencer Marketing Hub's report indicated that two Chinese digital influencers (i.e., Li Jiaqi and Viya) sold $3 billion worth of goods in one day (Geyser 2022).

On the other hand, unlike real humans, VIs are assets that can be manufactured, modified, bought, and sold (Robinson 2020; Drenten and Brooks 2020). These assets' owners are companies or actual people. According to Choudhry et al. (2022), among the 147 VIs worldwide, a single brand owns 23 and creates content to promote that brand's products exclusively, while the remainder collaborates with brands. Before discussing VIs and their effect on fashion marketing, it is necessary to understand the definitions of and distinctions between a digital influencer and a VI.

9.4.1 What Is a Digital Influencer?

Digital influencers (DIs) is a broader term used to identify those who produce online content not only for social media, but any digital platform, and they are not necessarily human beings. According to their follower numbers, DIs can be classified as a celebrity influencer (more than 1.5M followers), mega influencer (1.5M to 500K followers), macro influencer (500K to 100K followers), and micro-influencer (100K to 10K followers) (Metrics 2018). Using DIs in fashion marketing is inevitable today. According to Berezhna (2018), 98% of fashion brands had an Instagram account, and 90% of professionals in the marketing field believe that DIs play an important role in increased product awareness and communication with target consumers (Influencer MarketingHub 2019).

9.4.2 What Is a Virtual Influencer?

VIs are a subset of DIs. According to Arsenyan and Mirowska (2021), VIs are agents augmented with digital avatars, and as such, do not exist in the physical world. However, they differ from avatars because they are not the users' digital counterparts, but are more similar to human influencers, as they have their own character and communicate with their followers through their own stories (Hanus and Fox

2015). Although the majority of VIs look like real humans, for example, Lil Miquela, Shudu Gram, and Imma Gram (Choudhry et al. 2022), there are also doll-like VIs (e.g., Barbie, Noonoouri) and non-human-like VIs such as Janky, Guggimon, and GEICO's Gecko (VirtualHumans n.d.). Human-like VIs are so realistic that some followers do not even realize that they are a CGI. Storytelling is among the key aspects that make a virtual character successful in marketing (Moustakas et al. 2020), sometimes the VIs' creators prefer to veil the influencer's virtual nature, as is the case with Lil Miquela. Its creators posted an item on her Instagram account that declared that she was a virtual human only after she reached a considerable number of followers.

9.4.3 How Do VIs Influence People?

It is critical to understand the way VIs bond with their followers because eventually this bond increases or decreases followers' interest in a brand or a product (Jiménez-Castillo and Sánchez-Fernández 2019). The two-step flow of communication model by Katz and Lazarsfeld (1955) defines opinion leaders' influence within a community. According to the model, opinion leaders receive information (often from mass media) that they filter and develop before they transmit it to others via word of mouth (WOM). Therefore, they act as intermediaries between the information source and their followers. Similarly, DIs and VIs serve as online opinion leaders and distribute the filtered information via e-WOM (Hwang and Zhang 2018). Hennig-Thurau et al. (2004) described e-WOM as any statement about a company or a product that is distributed to a larger audience through online channels.

Although the two-step flow of communication model explains the way information is transmitted between influencers and their followers, it does not explain how exactly VIs influence people. To be influenced, people must accept the information they receive and learn from it so that they can internalize it. This condition requires the social learning theory to be considered (Bandura and Walters 1977), which suggests that people within a social network observe and imitate each other. In the social media context, DIs function as role models from whom people learn and who influence their attitudes and behaviours (Cheng et al. 2021). However, the influencers do not affect everyone equally. As the Media Dependency Theory states (Ball-Rokeach and DeFleur 1976), the more people are connected and dependent on social media, the more likely they are to adopt the information distributed and include it in their behaviours. Kadekova and Holienciova (2018) showed that age is an important factor in a DI's effect. Compared to Generation Z, Generation Y appears to be more suspicious about DIs' recommendations and less likely to purchase products that they endorse. Moreover, Hudders and De Jans (2022) discussed the effect of followers' and DIs' genders on the persuasiveness of an endorsed product. The results showed that DIs' gender was not important to persuasion directly; however, when the DIs and their followers were both women, the persuasion effect was greater, while male followers did not care whether the DI was a woman or a man. A DI's attractiveness is also key to understand the influence mechanism explained in detail in Torres, Augusto, and Matos's (2019) study, where attractiveness had a positive effect on influencing power. However, it should be noted that although physical attractiveness is a large component of attractiveness, other factors, such as behavioural attractiveness (i.e., sensuality, talent), are also important (da Silva Oliveira and Chimenti 2021). On the other hand, photographs are the key to drawing attention to an online post, particularly when the influencer's body is accentuated by tight clothing, bikinis, etc. Further, when the influencer her/himself is present in the post with the product being endorsed, it is more likely to engage people (Silva et al. 2020).

9.5 THE SOCIAL INTERACTION BETWEEN VIS AND REAL PEOPLE

The number of VIs on social media is increasing rapidly, as are their followers. The majority of followers begin to follow VIs through Instagram recommendations, e-WOM, or another influencer's reposts, while the stories and entertaining content that VIs share engage people and motivate them to follow VIs (Choudhry et al. 2022). The Influencer Marketing Factory surveyed more than 1,000 North Americans and found 58% of them followed at least one VI, while the follow rate was 75% for the 18–24 age group. The content, storytelling, aesthetics, and interaction are the main aspects of VIs or their posts that motivate people to follow them (The Influencer Marketing Factory 2022). Like human influencers, most of the content VIs share are about fashion, followed by music and art. The other reasons why people engage with VIs are their novelty and responsiveness.

9.5.1 Why Do People Interact with VIs?

VIs' novelty arouses people's curiosity and draws attention to the products they advertise. Although the Innovation Hypothesis can explain this curiosity (Kiesler and Sproull 1997), its effect vanishes when people become accustomed to it. Von der Pütten and colleagues (2010) summarized three other hypotheses and models (i.e., the Deficit Hypothesis, Ethopoeia Concept, and Threshold Model of Social Influence) that explain directly or indirectly why the interaction between a real human and a computer (in this case VIs as CGHs) occurs. The deficit hypothesis that Bernstein (1959) proposed first claims that people tend to interact with computers because of their inexperience. The Ethopoeia Concept suggests that social interaction occurs automatically in the presence of social cues, such as interaction through body language or speech, and whether this interaction is human–human or human–computer is unimportant (Nass and Moon 2000). The threshold model of social influence is more complex (Blascovich 2002); however, it can be simplified as social verification of a virtual communication in which the parties engage in a meaningful conversation that leads ultimately to a shared experience (Dodds and Watts 2009). On the other hand, the Parasocial Interaction (PSI) theory suggests that people are eager to develop one-sided feelings towards a media character, particularly if this media character responds to the follower (Horton and Wohl 1956). Generally, all communication is open to the public in social media, and therefore, there is an ongoing performance on the virtual stage. It is possible that VIs' responsiveness, particularly if they are human-like, creates a parasocial interaction between them and their followers (Choudhry et al. 2022).

9.5.2 How Do People React to VIs?

Research on social influence has shown that people's reaction to computers is nearly the same as that to real people (Nass and Moon 2000). More recent studies have narrowed the computer–human interaction to virtual agent–human interaction and found that people react the same to real humans and virtual agents (Von der Pütten et al. 2010; Krämer et al. 2015). Although people are willing to interact with VIs (Silva and Bonetti 2021), to what extent they accept them remains a question. Molin and Nordgren (2019) suggested that VIs are more likely to be accepted if they possess "humanness" (i.e., similarity to a real human), social interactivity (i.e., capability to interact socially with their followers), and social presence (i.e., availability).

Arsenyan and Mirowska's (2021) study questioned whether people's reactions to different kinds of influencers varied. They explored three types of DIs: a human influencer, human-like VI, and doll-like VI. According to the number of likes and video views, the study participants were most engaged with the human-like VI and least with real human influencers. However, participants had the most negative reactions to the Instagram posts of the human-like VI (Arsenyan and Mirowska 2021). The authors explained this attitude with the uncanny valley hypothesis, which states that nearly human-like entities make people feel uncomfortable (Mori et al. 2012). Although a VI's humanness increases people's acceptance of them (Molin and Nordgren 2019), excessive humanness may reduce parasocial interaction (Silva and Bonetti 2021).

9.6 VIRTUAL INFLUENCERS' PRODUCT ENDORSEMENT AND BRAND ADVERTISING

Even though, in the past five years, many fashion companies have collaborated with Vis, fashion companies hesitate employing VIs versus human influencers in terms of advertising. First of all, celebrity marketing is expensive for brands. For instance, one Instagram post by Christian Ronaldo costs one million dollars if it is a collaboration (Business of Apps 2022). On the other hand, virtual humans are less expensive. Once brands create their own virtual humans, they can use them as VIs constantly without extra costs. However, rather than a company creating its own VI, collaborating with a VI may be more expensive than collaborating with a human influencer (Baklanov 2019).

Another important aspect in influencer marketing is the congruence between the influencer and the brand or product endorsed. According to the Match-up Hypothesis (Kamins 1990), people are more likely to purchase a product when the influencer and the product endorsed fit (Till and Busler 2000). When companies create their own VIs, their congruence is not a problem, as the brand itself controls the VI's design, appearance, and online behaviours, according to their own expectations. Further, this freedom of control offers another advantage, which is minimizing a human influencer's PR risk. The Meaning Transfer Model (McCracken 1986) explains the way symbolic meanings are transferred from a celebrity to a brand and the way this influences people's interpretation of the brand or product that a celebrity is advertising. Consistent with the Meaning Transfer Model, an influencer's transgression may decrease consumers' trust in the brand and reduce their intention to purchase the products endorsed (Reinikainen et al. 2021).

VIs communicate with their followers either through AI or a real person at the keyboard. An AI influencer's transgression is perceived to be more acceptable and less likely to reduce trust in the brand (Thomas and Fowler 2021). On the other hand, VIs pose less PR risk if the brand itself controls them rather than collaborating with a VI that a third party controls. Finally, VIs can increase their followers' WOM intentions better than their real counterparts (Sands et al. 2022).

9.6.1 Challenges of VI Marketing

Despite the many advantages noted previously, VI marketing also has some challenges. These include maintaining the followers' engagement, as VIs' novelty naturally wanes over time when people become accustomed to them, as well as their authenticity (Moustakas et al. 2020). People require the influencer to be authentic and credible to be persuaded by them. Although VIs' human-like appearance can increase their trustworthiness, studies have shown that people trust VIs less than human influencers (Djafarova and Rushworth 2017). Similarly, market research has found that people trust the products VIs advertise only moderately (The Influencer Marketing Factory 2022). Although the report does not mention whether the difference between the age groups' trust was significant, the 35–45 age group demonstrated greater trust than the other age groups, including Generation Z, while the oldest group (55+) had the least trust.

Transparency is another issue that must be considered in VI marketing, as these influencers' virtuality is sometimes blurred, as in the case of Lil Miquela's early Instagram posts. Drenten and Brooks (2020) argued that transparency is not as critical as one would assume because the younger generation does not mind if the influencer is a virtual or real human. However, Fullscreen TBH Community's (2019) market report revealed that 42% of Generation Z and millennials wanted to know the company that controlled the VIs (Fullscreen TBH Community 2019). Finally, there is a debate about whether VIs can be held accountable for the products they endorse. Thomas and Fowler (2021) explained this according to the "Bona Fide User" Principle, which indicates that an influencer must recommend a product s/he endorses via a paid partnership honestly and not mislead the followers. However, VIs are not real. As a result, they cannot use a product or experience a service, and thus, they cannot recommend them. Hence, their credibility in this respect is questionable. Further, lawful VIs cannot be held accountable if the product they endorsed fails (Robinson 2020).

9.7 CONCLUSION

In the next decade, research related to the use of virtual humans is expected to increase for two reasons. The first is related to the fact that the COVID-19 pandemic changed the way we perceive the world. For example, distance education and remote work became norms. In the near future, we may accomplish many tasks with our avatars in the metaverse. The second reason is the fact that creating virtual humans is becoming easier because of the simplified 3D modelling software programs available today. Further, virtual humans' human-likeness is becoming flawless because of the use of AI. These technological advancements guarantee that we will encounter virtual humans not only in social media, but in other facets of our lives, such as online education, customer service, etc. Currently, avatars and the metaverse are still in their infancy. Nonetheless, virtual humans that are represented accurately can encourage people to attend these types of events and create potential marketing opportunities for fashion companies. In addition to the clear need to develop high-quality virtual humans, more work is required to understand the uncanny effect better and ways to overcome it to increase virtual humans' wide acceptance. Moreover, many people remain unaware that virtual humans exist (The Influencer Marketing Factory 2022). To make VI marketing strategies more effective and increase people's acceptance of them, awareness must be increased first, which can be accomplished by increasing VIs' presence on social media.

The rise of the metaverse is increasing the technological ventures in virtual humans and encouraging influencer marketers to use VIs. Complex 3D human modelling is increasingly becoming easier and less expensive, thus providing an opportunity for influencer marketers who hesitate to create their own VI. Fashion companies are now hiring Chief of Metaverse to explore and invest in the meta experience more often (McDowell 2021), and VI marketing can become one among Chief of Metaverse's responsibilities. In an era in which marketing's future is being questioned because big data are replacing marketers, it is important to follow the technological advancements in the field of AI and virtual humans and begin to invest in this relatively novel marketing strategy with VIs.

REFERENCES

Arsenyan, Jbid, and Agata Mirowska. 2021. "Almost human? A comparative case study on the social media presence of virtual influencers." *International Journal of Human-Computer Studies* 155 (July): 102694. https://doi.org/https://doi.org/10.1016/j.ijhcs.2021.102694.

Baek, Seung-Yeob, and Kunwoo Lee. 2012. "Parametric human body shape modeling framework for human-centered product design." *Computer-Aided Design* 44 (1): 56–67. https://doi.org/https://doi.org/10.1016/j.cad.2010.12.006.

Baklanov, N. 2019. "The top Instagram virtual influencers in 2019." *Hype-Journal*. https://doi.org/https://doi.org/10.1080/00913367.2020.1810595.

Ball-Rokeach, S.J., and M.L. DeFleur. 1976. "A dependency model of mass-media effects." *Communication research* 3 (1): 3–21. https://doi.org/https://doi.org/10.1177/009365027600300101.

Bandura, Albert, and Richard H. Walters. 1977. *Social Learning Theory*, Vol. 1. Englewood cliffs, NJ: Prentice Hall.

Bauld, Andrew. 2022. "What will learning in the metaverse look like? New guide gives educators a place to start figuring it out." www.gse.harvard.edu/news/uk/22/06/what-will-learning-metaverse-look.

Berezhna, V. 2018. "Brands boost influencer marketing budgets." *Business of Fashion* 6.

Bernstein, Basil. 1959. "A public language: Some sociological implications of a linguistic form." *The British Journal of Sociology* 10 (4): 311–326.

Blascovich, Jim. 2002. "Social influence within immersive virtual environments." In *The Social Life of Avatars*, 127–145. London: Springer.

Brown, Duncan, Nick Hayes, and Y.L. Chu. 2015. *Influencer Marketing: Who Really Influences Your Customers?* Amsterdam, Netherland: Elsevier.

Brownridge, Andrew, and Peter Twigg. 2014. "Body scanning for avatar production and animation." *International Journal of Fashion Design, Technology and Education* 7 (2): 125–132. https://doi.org/https://doi.org/10.1080/17543266.2014.923049.

Burns, K.S. 2021. "The history of social media influencers." In *Research Perspective on Social Media Influencers and Brand Communication*, edited by B. Watkins, 1–23. essay, London: Lexington Books.

Business of Apps. 2022. "Influencer marketing costs." *Featured Influencer Marketing Companies*. Business of Apps. Accessed July 18, 22. www.businessofapps.com/marketplace/influencer-marketing/research/influencer-marketing-costs/.

Casaló, Luis V., Carlos Flavián, and Sergio Ibáñez-Sánchez. 2020. "Influencers on Instagram: antecedents and consequences of opinion leadership." *Journal of Business Research* 117: 510–519. https://doi.org/https://doi.org/10.1016/j.jbusres.2018.07.005.

Cheng, Yang, Chun-Ju Flora Hung-Baesecke, and Yi-Ru Regina Chen. 2021. "Social media influencer effects on CSR communication: The role of influencer leadership in opinion and taste." *International Journal of Business Communication*: 2329488 4211035112.

Choudhry, Abhinav, Jinda Han, Xiaoyu Xu, and Yun Huang. 2022. "'I felt a little crazy following a 'Doll'' investigating real influence of virtual influencers on their followers." *Proceedings of the ACM on Human-Computer Interaction* 6 (GROUP): 1–28.

da Silva Oliveira, Antonio Batista, and Paula Chimenti. 2021. "'Humanized Robots': A proposition of categories to understand virtual influencers." *Australasian Journal of Information Systems* 25.

De Perthuis, Karen, and Rosie Findlay. 2019. "How fashion travels: The fashionable ideal in the age of Instagram." *Fashion theory* 23 (2): 219–242. https://doi.org/https://doi.org/10.1080/1362704X.2019.1567062.

Djafarova, Elmira, and Chloe Rushworth. 2017. "Exploring the credibility of online celebrities' Instagram profiles in influencing the purchase decisions of young female users." *Computers in Human Behavior* 68: 1–7. https://doi.org/https://doi.org/10.1016/j.chb.2016.11.009.

Dodds, Peter, and Duncan J. Watts. 2009. "Threshold models of social influence." In *The Oxford Handbook of Analytical Sociology*. Oxford: Oxford University Press.

Drenten, Jenna, and Gillian Brooks. 2020. "Celebrity 2.0: Lil Miquela and the rise of a virtual star system." *Feminist Media Studies* 20 (8): 1319–1323. https://doi.org/https://doi.org/10.1080/14680777.2020.1830927.

Farveen, Farzanah. 2020. "PUMA brings to life virtual influencer in SEA marketing push." *Marketing-Interactive*. Accessed July 17, 2022. www.marketing-interactive.com/puma-clicks-with-virtual-influencer-maya-for-sea-marketing-push.

Fullscreen TBH Community. 2019. "Bot or not?". Accessed July 18, 2022. https://grin.co/wp-content/uploads/2021/10/Fullscreen_CGI-Influencers_Bot-Or-Not.pdf.

Geyser, Werner. 2022. "The state of influencer marketing 2022: Benchmark report." *Influencer MarketingHub*. Accessed July 17, 2022. https://influencermarketinghub.com/influencer-marketing-benchmark-report/.

Goldstein, Michael S. 1992. *The Health Movement: Promoting Fitness in America*. Woodbridge: Twayne Publishers.

Gu, Liwen, Cynthia Istook, Yanwen Ruan, Godfree Gert, and Xiaogang Liu. 2019. "Customized 3D digital human model rebuilding by orthographic images-based modelling method through open-source software." *The Journal of The Textile Institute* 110 (5): 740–755. https://doi.org/https://doi.org/10.1080/00405000.2018.1548079.

Hanus, Michael D., and Jesse Fox. 2015. "Persuasive avatars: The effects of customizing a virtual salesperson' s appearance on brand liking and purchase intentions." *International Journal of Human-Computer Studies* 84: 33–40.

Hennig-Thurau, Thorsten, Kevin P. Gwinner, Gianfranco Walsh, and Dwayne D. Gremler. 2004. "Electronic word-of-mouth via consumer-opinion platforms: What motivates consumers to articulate themselves on the internet?" *Journal of Interactive Marketing* 18: 38–52. https://doi.org/https://doi.org/10.1002/dir.10073.

Hill, Kashmir, and Jeremy White. 2020. "Designed to deceive: Do these people look real to you?" *New York, NY: The New York Times*. Accessed July 15, 2022. www.nytimes.com/interactive/2020/11/21/science/artificial-intelligence-fake-people-faces.html.

Horton, Donald, and Richard Wohl R. 1956. "Mass communication and para-social interaction: Observations on intimacy at a distance." *Psychiatry* 19: 215–229.

Hudders, Liselot, and Steffi De Jans. 2022. "Gender effects in influencer marketing: An experimental study on the efficacy of endorsements by same-vs. other-gender social media influencers on Instagram." *International Journal of Advertising* 41 (1): 128–149.

Hwang, K., and Q. Zhang. 2018. "Influence of parasocial relationship between digital celebrities and their followers on followers' purchase and electronic word-of-mouth intentions, and persuasion knowledge." *Computers in Human Behavior* 87 (January): 155–173. https://doi.org/https://doi.org/10.1016/j.chb.2018.05.029.

The Influencer Marketing Factory. 2022. "Virtual influencers survey+ INFOGRAPHIC." Accessed July 17, 2022. https://theinfluencermarketingfactory.com/virtual-influencers-survey-infographic/.

Influencer MarketingHub. 2019. "Influencer marketing benchmark report: 2019."

Iqbal, Mansoor. 2022. "Instagram revenue and usage statistics (2022)." *Business of Apps*. Accessed July 17, 2022. www.businessofapps.com/data/instagram-statistics/.

Isaac, M. 2021. "Facebook changes corporate name to meta." *The New York Times*. Accessed July 17, 2022. www.nytimes.com/2021/10/28/technology/facebook-meta-name-change.html.

Jiang, Bo, Fukai Zhao, and Xinguo Liu. 2014. "Observation-oriented silhouette-aware fast full body tracking with Kinect." *Journal of Manufacturing Systems* 33 (1): 209–217. https://doi.org/https://doi.org/10.1016/j.jmsy.2013.10.003.

Jiménez-Castillo, David, and Raquel Sánchez-Fernández. 2019. "The role of digital influencers in brand recommendation: Examining their impact on engagement, expected value and purchase intention." *International Journal of Information Management* 49: 366–376. https://doi.org/https://doi.org/10.1016/j.ijinfomgt.2019.07.009.

Kadekova, Zdenka, and Mária Holienciova. 2018. "Influencer marketing as a modern phenomenon creating a new frontier of virtual opportunities." *Communication Today* 9(2): 90–10. https://communicationtoday.sk/download/22018/06.-KADEKOVA-HOLIENCINOVA-%25E2%2580%2593-CT-2-2018.pdf.

Kamins, Michael A. 1990. "An investigation into the 'match-up' hypothesis in celebrity advertising: When beauty may be only skin deep." *Journal of Advertising* 19: 4–13. https://doi.org/https://doi.org/10.1080/00913367.1990.10673175.

Katz, Elihu, and Paul F. Lazarsfeld. 1955. *Personal Influence: The Part Played by People in the Flow of Communication*. New York, NY: Free Press.: Routledge.

Kiesler, S., and L. Sproull. 1997. "Social human-computer interaction." In *Human Values and the Design of Computer Technology*. Cambridge: University Press.

Kim, Sung-Ho, and Kyung-Yong Chung. 2015. "Medical information service system based on human 3D anatomical model." *Multimedia Tools and Applications* 74 (20): 8939–8950.

Krämer, Nicole C., Astrid M. Von der Pütten, and Laura Hoffmann. 2015. "Social effects of virtual and robot companions." In *The Handbook of the Psychology of Communication Technology*, 137–159. Oxford: John Wiley & Sons.

Lee, Kyu Sun, and Hwa Kyung Song. 2021. "Automation of 3D average human body shape modeling using Rhino and Grasshopper Algorithm." *Fashion and Textiles* 8 (1): 1–20. https://doi.org/https://doi.org/10.1186/s40691-021-00249-6.

Li, Feng, and Timon C. Du. 2011. "Who is talking? An ontology-based opinion leader identification framework for word-of-mouth marketing in online social blogs." *Decision Support Systems* 51 (1): 190–197. https://doi.org/https://doi.org/10.1016/j.dss.2010.12.007.

Li, Zhen, Jianping Hao, and Cuijuan Gao. 2021. "Overview of research on virtual intelligent human modeling technology." 2021 IEEE Asia-Pacific Conference on Image Processing, Electronics and Computers (IPEC).

Magnenat-Thalmann, N. ed. 2010. *Modeling and Simulating Bodies and Garments*. Springer-London.

Myers, J. H., & Robertson, T. S. 1972. Dimensions of Opinion Leadership. *Journal of Marketing Research, 9*(1), 41–46.

McCann, Margaret, and Alexis Barlow. 2015. "Use and measurement of social media for SMEs." *Journal of Small Business and Enterprise Development* 22 (2): 273–287. https://doi.org/https://doi.org/10.1108/JSBED-08-2012-0096.

McCracken, Grant. 1986. "Culture and consumption: A theoretical account of the structure and movement of the cultural meaning of consumer goods." *Journal of Consumer Research* 13 (1): 71–84.

McDowell, Maghan. 2021. "Is it time to hire a chief metaverse officer?" *Vogue-Business*. Accessed July 17, 2022. www.voguebusiness.com/technology/is-it-time-to-hire-a-chief-metaverse-officer.

Metrics, Launch. 2018. "*The state of influencer marketing in fashion, luxury and cosmetics.*" *Launch Metrics*. Accessed January 29, 2023. https://www.launchmetrics.com/resources/blog/state-of-influencer-marketing-2018.

Molin, Victoria., and Sofia Nordgren. 2019. *Robot or human? The marketing phenomenon of virtual influencers: A case study about virtual influencers* (Master Thesis). Uppsala University—Department of Business studies. http://uu.diva-portal.org/smash/get/diva2:1334486/FULLTEXT01.pdf.

Mori, M., K.F. MacDorman, and N. Kageki. 2012. "The uncanny valley [From the field]." *IEEE Robotics & Automation Magazine*: 98–100.

Moustakas, Evangelos, Nishtha Lamba, Dina Mahmoud, and C. Ranganathan. 2020. "Blurring lines between fiction and reality: Perspectives of experts on marketing effectiveness of virtual influencers." *International Conference on Cyber Security and Protection of Digital Services (Cyber Security)*: 1–6. IEEE. https://doi.org/https://doi.org/10.1109/CyberSecurity49315.2020.9138861.

Nashville Film Institute. n.d. "What is CGI?—Everything you need to know." *Nashville Film Institute*. Accessed July 17, 2022. www.nfi.edu/what-is-cgi/.

Nass, C., and Y. Moon. 2000. "Machines and mindlessness: Social responses to computers." *Journal of Social Issues* 56 (1): 81–103.

Powers, Devon, and D.M. Greenwell. 2017. "Branded fitness: Exercise and promotional culture." *Journal of Consumer Culture* 17 (3): 523–541. https://doi.org/https://doi.org/10.1177/1469540515623606.

Rehak, Bob. 2011. "Computer-generated imagery." *Cinema and Media Studies*. https://doi.org/10.1093/OBO/9780199791286-0068.

Reinikainen, Hanna, Teck Ming Tan, Vilma Luoma-aho, and Jari Salo. 2021. "Making and breaking relationships on social media: The impacts of brand and influencer betrayals."

Technological Forecasting and Social Change 171: 120990. https://doi.org/https://doi.org/10.1016/j.techfore.2021.120990.

Robinson, Ben. 2020. "Towards an ontology and ethics of virtual influencers." *Australasian Journal of Information Systems* 24. https://doi.org/https://doi.org/10.3127/ajis.v24i0.2807.

Sands, Sean, Colin L. Campbell, Kirk Plangger, and Carla Ferraro. 2022. "Unreal influence: leveraging AI in influencer marketing." *European Journal of Marketing* 56 (6): 1721–1747.

Santiago, Joanna Krywalski, and Inês Moreira Castelo. 2020. "Digital influencers: An exploratory study of influencer marketing campaign process on Instagram." *Online Journal of Applied Knowledge Management (OJAKM)* 8 (2): 31–52. https://doi.org/https://doi.org/10.36965/OJAKM.2020.8(2)31-52.

Santora, Jacinda. 2022. "Key influencer marketing statistics you need to know for 2022." *Influencer MarketingHub*. Accessed July 18, 2022. https://influencermarketinghub.com/influencer-marketing-statistics/.

Silva, Emmanuel Sirimal, and Francesca Bonetti. 2021. "Digital humans in fashion: Will consumers interact?" *Journal of Retailing and Consumer Services* 60: 102430.

Silva, Marianny Jessica de Brito, Salomão Alencar de Farias, Michelle Kovacs Grigg, and Maria de Lourdes de Azevedo Barbosa. 2020. "Online engagement and the role of digital influencers in product endorsement on Instagram." *Journal of Relationship Marketing* 19 (2): 133–163. https://doi.org/https://doi.org/10.1080/15332667.2019.1664872.

Sturman, David J. 1994. "A brief history of motion capture for computer character animation." *SIGGRAPH94, Course9*.

Thomas, Veronica L., and Kendra Fowler. 2021. "Close encounters of the AI kind: Use of AI influencers as Brand endorsers." *Journal of Advertising* 50 (1): 11–25.

Till, Brian D., and Michael Busler. 2000. "The match-up hypothesis: Physical attractiveness, expertise, and the role of fit on brand attitude, purchase intent and brand beliefs." *Journal of Advertising* 29 (3): 1–13. https://doi.org/https://doi.org/10.1080/00913367.2000.10673613.

Time staff. 2017. "The 25 most influential people on the internet." *Time Magazine*. Accessed January 29, 2023. https://time.com/4815217/most-influential-people-internet/.

Torres, Pedro, Mário Augusto, and Marta Matos. 2019. "Antecedents and outcomes of digital influencer endorsement: An exploratory study." *Psychology & Marketing* 36 (12): 1267–1276.

Uzunoğlu, Ebru, and Sema Misci Kip. 2014. "Brand communication through digital influencers: Leveraging blogger engagement." *International Journal of Information Management* 34 (5): 592–602. https://doi.org/https://doi.org/10.1016/j.ijinfomgt.2014.04.007.

VirtualHumans. 2020. "5 notable virtual influencer + brand partnerships." Accessed July 16, 2022. www.virtualhumans.org/article/5-notable-virtual-influencer-brand-partnerships.

VirtualHumans. 2021a. "Prada creates virtual muse named candy." Accessed July 17, 2022. www.virtualhumans.org/article/prada-creates-first-virtual-muse-candy.

VirtualHumans. 2021b. "Welcoming 27 newfound virtual influencers to VirtualHumans.org." www.virtualhumans.org/article/welcoming-29-newfound-virtual-influencers-to-virtualhumans-org.

VirtualHumans. 2021c. "What's the difference between virtual influencers, VTubers, artificial intelligence, avatars, and more?". Accessed July 16, 2022. www.virtualhumans.org/article/whats-the-difference-between-virtual-influencers-vtubers-artificial-intelligence-avatars.

VirtualHumans. n.d. "Welcoming 27 newfound virtual influencers to VirtualHimans.org." *Virtual influencers*. www.virtualhumans.org/#influencers.

Von der Pütten, Astrid M., Nicole C. Krämer, Jonathan Gratch, and Sin-Hwa Kang. 2010. "'It doesn't matter what you are!' Explaining social effects of agents and avatars." *Computers in Human Behavior* 26(6): 1641–1650. https://doi.org/10.1016/j.chb.2010.06.012.

Westerlund, Mika. 2019. "The emergence of deepfake technology: A review." *Technology Innovation Management Review* 9 (11). https://timreview.ca/article/1282.

Zhu, Shuaiyin, Pik Yin Mok, and Tsz-Ho Kwok. 2013. "An efficient human model customization method based on orthogonal-view monocular photos." *Computer-Aided Design* 45 (11): 1314–1332. https://doi.org/https://doi.org/10.1016/j.cad.2013.06.001.

Zhu, Wenmin, Xiumin Fan, and Yanxin Zhang. 2019. "Applications and research trends of digital human models in the manufacturing industry." *Virtual Reality & Intelligent Hardware* 1 (6): 558–579. https://doi.org/10.1016/j.vrih.2019.09.005.

10 A Review of Blockchain Technology for Sustainable Fashion

Sin Ying NG and P.Y. Mok

CONTENTS

10.1 INTRODUCTION

Consumer demands for sustainability in the fashion industry are soaring. Consumers across the globe have proactively participated in various worldwide campaigns for sustainable fashion with environmental NGOs to express their concerns about the lives of garment workers and transparency within the fashion industry. Who Made My Clothes, which was launched after the Rana Plaza tragedy in 2013, is one notable movement. The movement has encouraged millions of consumers to incorporate the #WhoMadeMyClothes hashtag and the name of the fashion brand that they are asking in social media posts. Fashion brands have been driven to respond to the sustainability issue to meet the needs of consumers.

However, there is no standard definition of sustainable fashion. Brands often used sustainable fashion, ethical fashion, slow fashion, green-fashion and eco-fashion

DOI: 10.1201/9781003264958-14

interchangeably, yet their meanings are different (Carey and Cervellon 2014; Henninger et al. 2016). Mukendi et al. (2020) considered sustainable fashion to be more related to ethical fashion, which includes social and environmental perspectives in various sectors. The sectors include: traceability (Henninger 2015; Kumar et al. 2017), ethical sourcing, environmentally friendly materials, circularity (Niinimäki 2018, 17), waste and product footprint, fair trade and the rights of garment workers (Henninger et al. 2016).

In order to drive sustainability, fashion brands leverage digital technology, such as 3D virtual prototyping, artificial intelligence (AI) and blockchain technology. 3D virtual prototyping technology has been adopted by leading brands aiming to reduce physical apparel prototypes and the environmental footprint in the product development process (Papahristou and Bilalis 2016; Papahristou and Bilalis 2017). The adoption of AI in fashion trend forecasting and supply chain management enables accuracy of sales demand prediction and real-time data, so decreasing overproduction and waste (Ivanov et al. 2019). Blockchain technology, which has been extended to fashion supply chain traceability since 2017, is another notable emerging technology. Its decentralized, timestamped and immutable features enable the tracking and tracing of fashion products from raw materials to finished garments in order to address a key pain point of fashion supply chain—a lack of supply chain transparency. In 2017, independent London designer Martine Jarlgaard collaborated with Provenance, a transparent company and fashion innovation agency, and presented the first ever blockchain proof-of-concept (POC) in the fashion industry at the Copenhagen Fashion Summit's Solution Lab (Beckwith 2018). The innovation enabled consumers to track and trace the entire supply chain journey of Martine Jarlgaard's knitwear product from farm to studio house. Figure 10.1 shows a near-field communication (NFC) chip attached to the label of a knitted jacket. Consumers could scan the NFC chip to access information, including location mapping, content and timestamps from raw materials to the finished garment. After the success of this POC, fashion brands in different market positions began to leverage blockchain technology for supply chain traceability and further expand to other areas of application in sustainable fashion.

This chapter aims to review the blockchain technology, the key areas of applications in sustainable fashion and the challenges of blockchain implementation. Academic and practitioner literature were reviewed in order to give a holistic perspective of research innovation and the practical use of blockchain technology for sustainable fashion. Web of Science and Scopus were chosen as the databases for searching academic literature. Women's Wear Daily (WWD), Business of Fashion (BOF) and Vogue Business were chosen as the sources for searching practitioner literature because they are leading publications and authorities in the fashion industry. The publications by blockchain service providers, such as IBM, VeChain, Textile Genesis and Everledger, were also reviewed. Table 10.1 highlights the search criteria. After screening, a total of 54 texts related to blockchain and sustainable fashion were found.

The rest of the chapter is structured as follows. Section 10.2 reviews the blockchain technology and explains why blockchain technology should be adopted for sustainable fashion. Four key areas of blockchain applications in sustainable fashion

FIGURE 10.1 Martine Jarlgaard's blockchain application to let consumers access the product manufacturing story.

Source: Arthur (2017).

TABLE 10.1
The Selection Criteria for Academic and Practitioner Literature

	Academic Literature	Practitioner Literature
Search sources	Web of Science and Scopus	Women's Wear Daily (WWD), Vogue Business, Business of Fashion (BOF) and publications by blockchain service providers
Keyword	"Blockchain or Distributed Ledger Technology", "Sustainability" and "Clothing, Textiles, Fashion or Apparel"	
Language	English	
Year	From the beginning of 2015 to Sep 2021	
Selection criteria	• No duplication • The focus of the literature should be on blockchain and sustainable fashion	
Number of texts after screening	27	27

are identified in section 10.3. Section 10.4 presents an in-depth evaluation of blockchain technology for sustainability in the fashion industry and highlights the challenges of implementation. Finally, Section 10.5 offers feasible suggestions to address the challenges of blockchain adoption.

10.2 HOW BLOCKCHAIN TECHNOLOGY COULD MAKE FASHION MORE SUSTAINABLE

Blockchain, as its name implies, is a shared digital ledger which includes a list of connected blocks stored on a distributed network. The blocks, containing metadata, are linked via cryptography to form a chain (Rauchs et al. 2018, 12). The data record is timestamped and cannot be altered. Though the tamper-proof concept of blockchain was introduced by Stuart Haber and W. Scott Stornetta in 1991, it took almost two decades before the first ever application of blockchain, Bitcoin, was launched in January 2009. This peer-to-peer cryptocurrency demonstrates several unique characteristics of blockchain technology, including *decentralization*, *immutability*, *consensus* and *provenance* (Andolfatto 2018; Kassab et al. 2019; Kassab 2021).

10.2.1 KEY CHARACTERISTICS OF BLOCKCHAIN TECHNOLOGY

The *decentralized* network enables every party in the network to have an exact same copy of the distributed ledger, which records and captures near real-time transactions from the beginning without a third-party intermediary. The transaction data in blockchain are secured by cryptographic methods such as a hashing algorithm, which turns data into a string of unique characters and is difficult to reverse or decrypt. Therefore, any transaction data written into the block is *immutable*, which means no one can change or tamper with the data (Kassab et al. 2019). Moreover, the transaction data should be verified and agreed by all parties in the network through a *consensus* mechanism before the new block is added to the blockchain (Kassab et al. 2019). The distributed ledger will be updated accordingly. If a party's ledger is altered and tampered with, it will be rejected by the majority of parties in the network. Additionally, the *provenance* feature allows authorized parties in the network to trace the origin of digital assets and the entire history of related transactions, so to increase transparency (Andolfatto 2018). These essential features make blockchain a strong cybersecurity technology to minimize fraud and enhance accurate tracking and the transparency of end-to-end tracking. Because of these benefits, blockchain has been widely applied in various industrial sectors not limited to cryptocurrency, such as trade finance, food, healthcare and supply chain management.

10.2.2 BLOCKCHAIN AND SUSTAINABILITY

Although the characteristics of blockchain enable digital transformation in various industries, the technology itself has been criticized by researchers as environmentally unsustainable (de Vries 2018; Howson and de Vries 2022; Sutherland 2019). Bitcoin is one of the most renowned examples of peer-to-peer cryptocurrency and will be issued to reward the "miners" when they have successfully solved the cryptographic puzzle and added a new block to the chain. This proof-of-work (PoW) mining process produces huge energy waste because, in order to compete with other miners, miners need powerful mining hardware and cooling system, which may result in higher electricity consumption and carbon footprint. According to Howson and de Vries (2022), up to November 2021, the estimated annual electricity consumption of

Bitcoin was 190 TWh, which is equal to the annual energy consumption of Thailand with its population of 69 million.

To address the huge energy consumption problem, proof-of-stake (POS), an energy efficient alternative, has been proposed and will be adopted in Ethereum blockchain very soon (Schinckus 2020). This mechanism allows the miner who has more assets to have a higher probability of validating a new block and receiving network fees as the reward. Hence, the miner will purchase more coins to become a bigger stakeholder instead of investing in computing power. Theoretically, this reduces the consumption of energy.

Direct acyclic graph (DAG)-based protocols were introduced; this is regarded as energy efficient and a greener option than the traditional blockchain data structure. Kotilevets et al. (2018) explain that DAG is a topological ordering structure in which blocks are not arranged sequentially one by one but can be arranged as multiple parallel chains in one direction, from earlier to later. That means that several transactions can be confirmed at the same time to increase the speed of transactions, thereby allowing transactions to be processed and confirmed efficiently. Figure 10.2 shows the structural difference between blockchain and DAC. IOTA is a software that adopts the DAG protocol and is widely known for its low energy consumption of 0.00016 kilowatt hours (kWh) for each transaction (Sori et al. 2021). That is significantly lower than the average energy consumption per transaction of MasterCard (0.0007 kWh) and Bitcoin, which exceeded 1,700 kWh by the end of 2018 (Sori et al. 2021; de Vries 2018). IOTA has collaborated with various companies, including

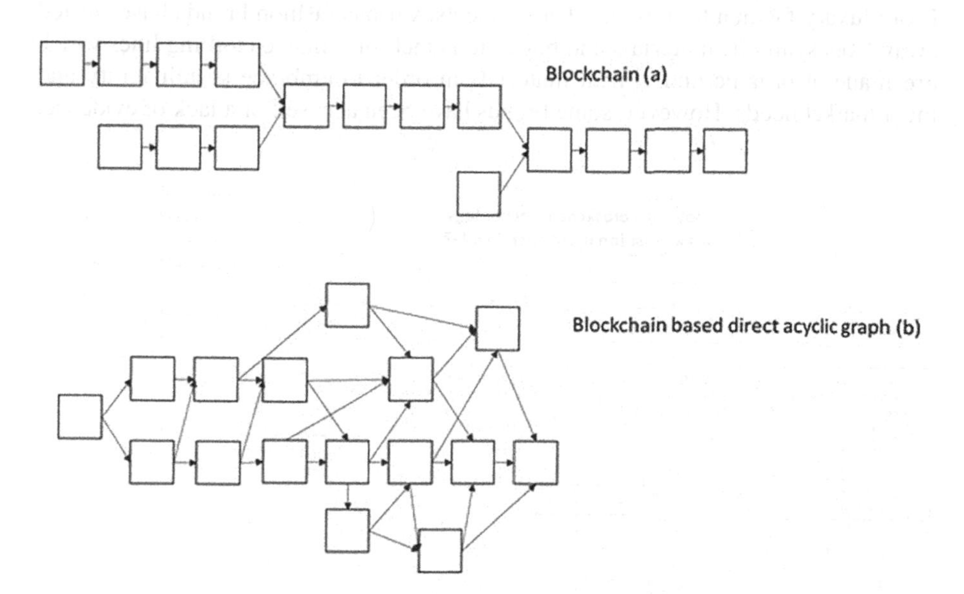

FIGURE 10.2 Structure of blockchain and direct acyclic graphic (DAG). (a) Blockchain is a shared digital ledger, which contains a list of connected blocks, linked via cryptography to form a chain. (b) DAG is a topological ordering structure in which blocks are chronologically arranged as multiple parallel chains.

ALYX, EVERYTHING and Avery Dennison, to achieve sustainable development goals. The team developed an end-to-end provenance solution to trace the journey of an ALYX garment from the raw materials to the display rack (Maguire 2019; IOTA Foundation 2019). Sustainability credentials are provided to consumers, while brands can have full visibility of their products throughout the supply chain.

The decentralized, immutable, consensus-based and provenance characteristics of blockchain technology enable different aspects of its application in sustainable fashion, which will be presented and discussed in the next sections.

10.3 ASPECTS OF BLOCKCHAIN TECHNOLOGY APPLICATION IN SUSTAINABLE FASHION

After reviewing the literature, four major aspects of blockchain technology in sustainable fashion are identified and discussed in this section. Figure 10.3 shows how the characteristics of blockchain enable sustainable fashion. The decentralization, immutability, consensus-based and provenance characteristics of blockchain technology provide proof of sustainable origin and embrace life cycle analysis, circular fashion, and workers' wellbeing and social welfare in sustainable fashion. The technology provides evidence and enhances transparency for sustainability compliance, and avoids greenwashing.

10.3.1 Proof of Sustainable Origin

From luxury fashion to mass market segments, various fashion brands have shifted away from synthetic materials and begun to launch sustainable clothing lines which are made of organic and ethical materials in order to embrace sustainability and meet market needs. However, some brands have been accused of a lack of evidence,

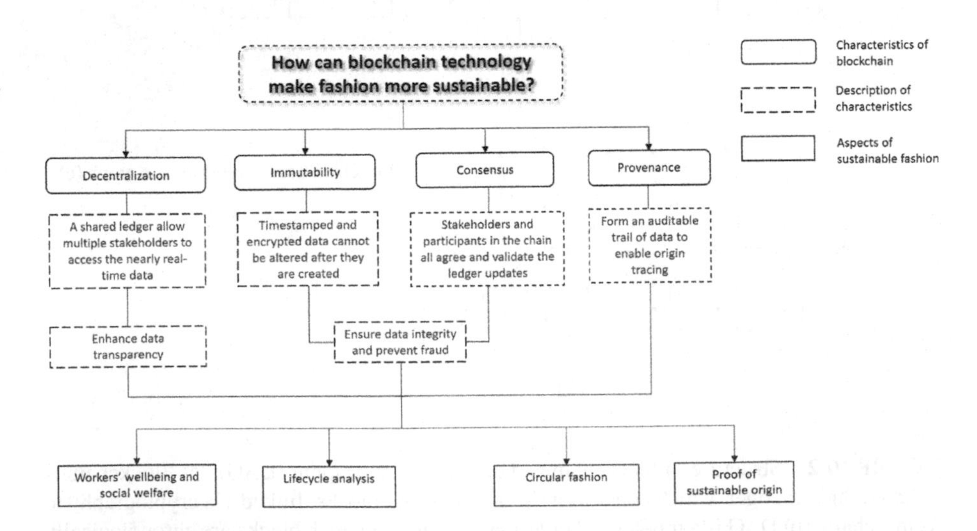

FIGURE 10.3 Mapping of blockchain application in various aspects of sustainable fashion.

transparency and verification of the product origin. To bridge the trust gap with customers, fashion companies, including material suppliers and fashion brands, leverage blockchain technology to verify the origin of their products. Since the first ever blockchain proof-of-concept in the fashion industry in 2017 at the Copenhagen Fashion Summit's Solution Lab, the fashion industry has collaborated with blockchain technology companies to track and trace the origin of sustainable materials.

Tokenization is a commonly used approach in the fashion industry to trace the origin of materials. Lenzing Group, a world-leading supplier of viscose fiber, collaborated with blockchain tech company Textile Genesis to conduct various rollouts to trace the entire product journey from fiber to the finished product through tokenization (BOF STUDIO 2021). Lenzing leveraged FiberCoins, blockchain-based digital tokens, which serve as fingerprints to represent the quantity of physical fiber shipment from the fiber supplier (i.e., each kilogram of fiber is represented in the platform as one FiberCoin). Only the fiber supplier (Lenzing) can issue FiberCoins and transfer coins in the platform to its downstream partners (e.g., spinners, fabric mills, dyeing houses, garment manufacturers and retailers) to trace and track the entire textile supply chain. The first rollout was conducted in 2019; Lenzing collaborated with WWF, Textile Genesis and Chicks to trace approximately 25,000 Chicks' products made of TENCEL™ fiber from fibers to retail (Lenzing May 2019; Lenzing Sep 2019). After the success of the first rollout, Lenzing and Textile Genesis started another 12-month pilot programme in November 2020 with H&M, Armed Angles, Mara Hoffman and Chicks to trace TENCEL™ and LENZING™ ECOVERO™ fibers (Lenzing 2020). It was estimated that most of the supply chain partners would onboard in Q2 of 2021. The platform enables brands and retailers to have full supply chain visibility of Lenzing's fiber and to access the digitally signed fabric certificates. In addition, other firms, such as Fashion for good, Kering's, Zalando, CanopyStyle and U.S. Cotton Trust Protocol, have partnered with Textile Genesis to trace sustainable viscose and cotton in industry-wide implementation (Fashion for good 2020; Carr 2020; Tenn 2021).

Scholars proposed a similar tokenization approach in the blockchain-based traceability system (Westerkamp et al. 2020; Agrawal et al. 2021). Westerkamp et al. (2020) proposed the issuance of a non-fungible digital token, which is a unique computer code representing a batch of goods, and the token can be transferred, split, merged or transferred to other partners within the supply chain system. Agrawal et al. (2021) proposed a private blockchain-based traceability framework with smart contracts and transaction rules with five main users, including organic cotton producers or suppliers, yarn manufacturers, fabric manufacturers, apparel manufacturers and retailers to track the organic cotton in the supply chain. Only the organic cotton producer is allowed to add organic cotton mass to the blockchain, representing the quantity of organic cotton produced, while other users are only allowed to transfer the mass to downstream supply chain partners. A smart contract is a computer programme that runs on a blockchain network to execute the contract or agreement automatically when predetermined conditions are met.

To further enhance the traceability of blockchain systems, internet of things (IoT) devices with embedded sensors, such as RFID tags, QR code and NFC chip, have been implemented to collect and transfer data over networks. The recycled wool collection of COS, a subsidiary brand of H&M, attached a label tag with a QR code on garments to let consumers scan the code and access information about the product, from location mapping to content and timestamps regarding the origin of the recycled wool (Vechain101 2020; DNV n.d.). Figure 10.4 shows the detail of a COS label with a QR code. Other fashion brands such as Arket and Martine Jarlgaard have used NFC technologies in their collections (Vechain101 2020; Beckwith 2018). The NFC chip is attached on to the labels of some selected fashion collections. Consumers can scan the NFC label on the physical garment to track and trace the life cycle of the product. An electronic product code (EPC) on a RFID tag is also proposed by some scholars to provide a globally unique object ID for every product and track the process of each product (Faridi et al. 2021; Pal and Yasar 2020). These technologies enhance the transparency and traceability of the product origin and provide visible evidence to support the sustainability claims.

10.3.2 Lifecycle Analysis and Emission Trading

In addition to transparency of the material origin, the environmental impact of a product through its lifecycle, from upstream to downstream of the fashion supply chain, is another major concern for sustainable fashion. A blockchain-enabled lifecycle assessment (LCA) framework has been proposed by scholars to obtain reliable tracking footprint and emissions data from the complex and fragmented supply chain (Kouhizadeh and Sarkis 2018; Zhang et al. 2020; the et al. 2020). Zhang et al.

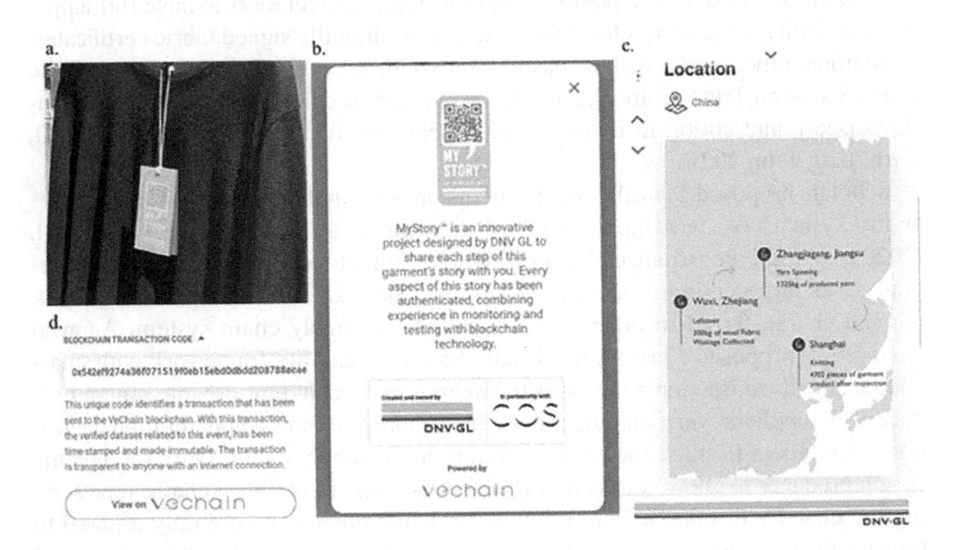

FIGURE 10.4 Example of leveraging label on a garment to let consumer access product information: (a) garment with "MyStory" label, (b) overview of "MyStory" label with a QR code, (c) product information displayed on a mobile phone after scanning the QR code and (d) the unique blockchain transaction code of product.

Source: VeChain 101 (2020).

(2020) proposed an LCA system architecture for various industries by enabling the sharing of data on input substances, waste, energy consumption and output emission to a wide range of stakeholders such as governments and enterprises. Enterprises could utilize the analyzed data to re-design products with smaller footprints. Teh et al. (2020) also highlighted that the use of blockchain technology for LCA in the material industries, including forest products, natural textiles and animal fibers, can provide accurate production data in real time from all suppliers, rather than an estimated amount in the current LCA protocol.

In addition, the industry shares the same vision as the academics. Covalent, the luxury brand renowned for bags and accessories made of carbon-negative bioplastic material "AirCarbon", collaborated with IBM and LinuxONE to track and disclose its carbon impact throughout the production process, by using blockchain technology to address consumer scepticism (Roshitsh 2020; IBM Newsroom 2021). Consumers access the Covalent blockchain webpage and enter the unique 12-digit code printed on the purchased item. The digit code, known as the Carbon Date, indicates the time the product was created by "AirCarbon". Consumers can trace the process of creation with a timestamp and the exact carbon footprints of the purchased product with a lifecycle analysis report verified by a third-party carbon accounting firm based on the ISO 14044:2006 standard—lifecycle assessment. Figure 10.5 and Figure 10.6 demonstrate the Covalent blockchain webpage and lifecycle analysis report.

In addition to carbon tracking, blockchain technology can even serve as an emissions trading tool, allowing firms to trade carbon credits. Usually one credit permits the emission of one tonne of carbon dioxide; therefore, firms that cut their carbon emissions can sell or trade their unused credits to others. This approach provides an economic incentive to firms to reduce emissions. However, the traditional process of calculating trading credits and emissions records can be time-consuming. With the blockchain ledger, the emissions data in different stages of production can be traced and recorded efficiently. Lam and Lei (2019) illustrated a case study of a

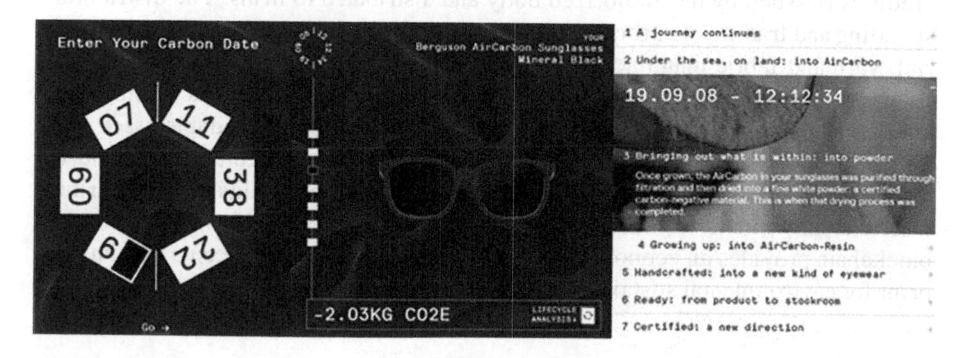

FIGURE 10.5 The Covalent blockchain webpage which enables users to enter the 12-digit "Carbon Date" code to discover the product's impact during the manufacturing processes. The exact time of each process is shown.

Source: Covalent (n.d.).

SCS Global Services does hereby certify that an independent assessment has been conducted on behalf of:

Newlight Technologies Inc

14382 Astronautics Dr, Huntington, CA, United States

For the following product(s):

Covalent AirCarbon™ Berguson Sunglasses

The product(s) meets all of the necessary qualifications to be certified for the following claim:

-2.03 kg CO2e

VERIFIED by SCS GLOBAL SERVICES CARBON FOOTPRINT

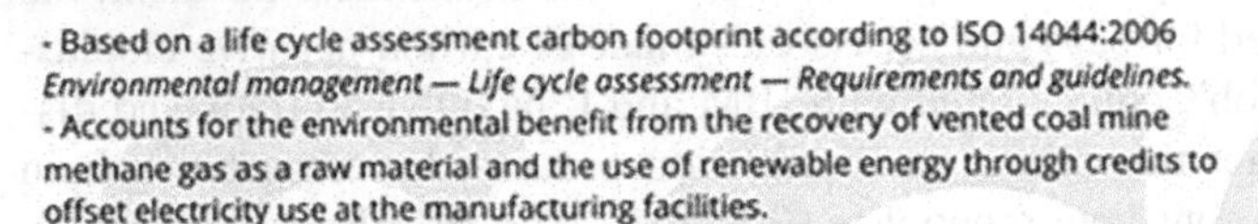

- Based on a life cycle assessment carbon footprint according to ISO 14044:2006 *Environmental management — Life cycle assessment — Requirements and guidelines.*
- Accounts for the environmental benefit from the recovery of vented coal mine methane gas as a raw material and the use of renewable energy through credits to offset electricity use at the manufacturing facilities.

Registration # SCS-CFP-00034

Valid from: September 22, 2020 to September 21, 2021

Stanley Mathuram, PE, Vice President

SCS Global Services

2000 Powell Street, Ste. 600, Emeryville, CA 94608 USA

FIGURE 10.6 The lifecycle assessment carbon footprint verification.

Source: Covalent (n.d.).

blockchain-based carbon footprint tracking system with an emissions trading feature for the textile and apparel supply chain which was jointly developed by Hong Kong Applied Science and Technology Research Institute and World Wide Fund for Nature (WWF) Hong Kong. A carbon token, a unique code to represent the carbon credit, was issued by the authorized body and distributed to firms. The distribution, spending and trading of carbon tokens in the production stages, including raw material, yarn and fabric manufacturing, garment sewing and finishing, and packaging processes, are recorded in the shared blockchain ledger, and this accounting tool enables automatic carbon footprint calculation and serves as a tradable asset. Fu et al. (2018) proposed a carbon credit-trading blockchain architecture to extend the trading to individual customers. Customers can gain carbon credits if they distribute their old garments to fashion retailers for recycling (Fu et al. 2018). The proposed blockchain provides an economic incentive to the industry to reduce its carbon footprint for environmental sustainability.

10.3.3 Circular Fashion

Besides emissions trading, the implementation of circular economy strategies can reduce carbon footprint and waste. According to Kouhizadeh and Sarkis (2018) and Reike et al. (2018), circular economy strategies in the fashion industry usually involve recycling, reuse and refurbishment. To embrace circularity and sustainability, fashion brands, from luxury fashion to mass market segments, have implemented related strategies, such as recycling and donating old clothes, reselling or swapping second-hand products, and clothing rental. However, some common obstacles have been found, including the lack of traceability of product origin information (i.e., quality assurance of material and production information) and lack of a standard policy, which might hinder the recycling and reselling process. To overcome these barriers, fashion brands leverage blockchain technology to trace the product origin for authenticity verification of pre-loved items.

The luxury fashion group LVMH introduced blockchain technology in pilot studies with blockchain technology company ConsenSys and Microsoft, forming a luxury goods tracking platform "AURA" in May 2019 (McDowell 2021b). It was an Ethereum consortium model with lists of luxury brands members such as Louis Vuitton and Christian Dior. It aimed to track the product lifecycle of luxury goods, from sourcing to sales, rental service and peer-to-peer second-hand marketplaces across the entire luxury industry.

Another luxury fashion blockchain example, the Arianne protocol, is a public and open source blockchain protocol that allows anyone to join, contribute and input information into the platform (McDowell 2019a; McDowell 2021b; Lee 2020). The product owner and brands can register the product on the platform by creating a non-fungible token (NFT) with a unique Arianee ID, which is a digital passport for a physical luxury object with information on proof of ownership, origin of garment, transfer capabilities, repair history and request, and authentication. The use of NFT as digital certificate streamlines the process of fashion circularity.

MCQ, a lower-priced diffusion line of luxury brand Alexander McQueen, collaborated with Everledger and Temera to develop "MyMCQ" (So 2020) (see Figure 10.7). Scanning the NFC label attached to a physical garment has two purposes: (1) it allows the consumer to track and trace the entire lifecycle of each MCQ item for authenticity verification, and (2) it allows the consumer to add transaction information to the blockchain to resell pre-owned products in a peer-to-peer second-hand marketplace.

Besides luxury fashion brands, fashion brands for the mass market also leverage blockchain to implement a circularity solution to embrace sustainability. H&M, PVH Corp and Target partnered with Eton, Microsoft and Waste Management to establish the CircularID initiative in 2019 to provide digital identification of physical apparel products (McDowell 2019b). The protocol aims to achieve two primary goals: (1) to identify products for rental and resale and (2) to identify the material components for product repair (e.g., identifying the lost button on a jacket in order to order and replace the exact button to repair the jacket) to lengthen the product life and enhance material recycling.

In addition to the finished product circularity, circularity in materials, such as reselling or reusing deadstock fabric, is a feasible attempt to achieve circular sustainability. Queen of Raw, a blockchain and machine learning-based marketplace for deadstock material, introduced MateriaMX to identify, track and trace deadstock fabric within supply chains in real-time, and automatically send it to the Queen of

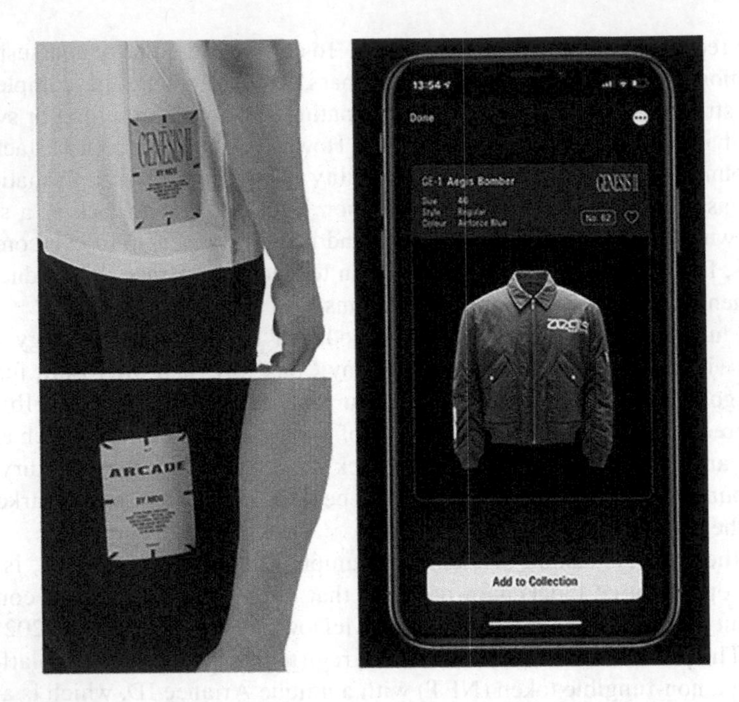

FIGURE 10.7 Left: the NFC is inserted inside the sewn label of MCQ garments. Right: The user interface demo of MyMCQ blockchain platform.

Source: So (2020).

Raw marketplace for resale (McDowell 2021a). A data report, including emission and environmental impact, is provided.

Scholars integrated the circularity in products and materials into a comprehensive blockchain framework for the circular fashion industry (Wang et al. 2020; Huynh 2021; Rusinek et al. 2018). Wang et al. (2020) designed a blockchain-based circular supply chain management system for the fast fashion sector, which covers various stages of fashion product lifecycle, from product design to after-sales service, sustainability assessment and material reuse. It is worth noting that the agile product design layer ensures products keep up with the fashion trend and enables efficient data sharing (e.g., change of design after receiving comments from clients) throughout the whole product lifecycle for fast fashion. Besides the fast fashion sector, Huynh (2021) proposed the adoption of blockchain in circular business models for the whole fashion industry. Firstly, a blockchain-based supply chain model is proposed to track product lifecycle and provide sustainability metrics of products to customers. Secondly, a service-based model aims (1) to offer clothing rental service and preloved product selling and repairing and (2) to leverage blockchain technology to increase sorting automation and recycling efficiency. The product lifecycle information (e.g., material composition, dyes and finishing process) stored in the blockchain can speed up the sorting process. In addition to these aspects, Rusinek et al. (2018) include long-term policy and management in their blockchain framework model because of the increasing compliance, traceability and data security pressure from policy makers. One example is the Circular Economy Package (CEP) in Europe, which includes measures and legislative proposals on waste; enterprises are required to follow the action plan which requires the re-use and recycling of municipal waste (i.e., textile waste) to at least 60%. CEP indicated the use of blockchain technology for close-loop supply chain management, enabling textile reuse and remanufacturing in order to address the regulatory requirement and governance for fashion circularity.

10.3.4 Workers' Wellbeing and Social Welfare

From the social perspective, poor working conditions and unfair treatment of workers are long-term issues in the labour-intensive garment industry, including long working hours, child labour, wage exploitation and labour abuse. Since the Rana Plaza tragedy in 2013, fashion brands have begun to improve employees' welfare and drive social sustainability in their vendor factories in order to respond to the surging demands from customers, trade unions, labour right NGOs and Corporate Social Responsibility (CSR) compliance.

From 2019 to 2020, Levi Strauss & Co., a fashion brand widely known for denim jeans, partnered with Harvard's T.H. Chan School of Public Health, the New America Foundation and ConSensys to develop an open-sourced blockchain platform "Survey Assure" that empowers approximately 5,000 workers in Mexican factories to express their daily experience and wellbeing in the workplaces, under the Sustainability and Health Initiative for NetPositive Enterprise (SHINE) programme (McDowell 2019a). Before the implementation of blockchain, the programme encountered the following pain points: (1) insecurity and the risk of survey data manipulation, (2) lack of transparency of survey results, and (3) the lengthy time (three months) to collect results, which makes it difficult to have effective and immediate responses (New America Foundation n.d.). Following the implementation of the "Survey Assure" platform, workers were invited to fill in the survey in Qualtrics anonymously with corresponding survey ID. The Ethereum-based blockchain platform consolidated survey responses from Qualtrics nearly real time, with automatically formatted data without manipulation and tampering presented visually in read-only charts, which would be accessible to both management teams and factory workers on mobile and screens (New America Foundation n.d.).

In addition to tracking workers' wellbeing, scholars have conducted mathematical modelling to show that the implementation of blockchain, government sponsors and an environmental taxation waiving scheme can enhance social welfare and alleviate the decrease in fashion supply chain profit caused by poor data quality issues in emerging markets (Choi and Luo 2019).

Table 10.2 and Table 10.3 summarize all the academic literature and industrial cases based on the four sectors of sustainable fashion. The benefits of incorporating blockchain technology are also highlighted.

TABLE 10.2
List of Papers and Benefits of Blockchain Technology

Aspects	Subdivision	References	Paper type	Benefits
Proof of sustainable origin	Fashion and textiles	Agrawal et al. 2018	Case study	**(1) Information transparency** The data are shared to all relevant parties in the blockchain ledger, including stakeholders, firms, brands, the government and customers to track the product origin. **(2) Instant traceability** Blockchain documents the provenance of products from the raw material to finished product in store. It enhances the supply chain visibility.
		Agrawal et al. 2021; Elmessiry and Elmessiry 2018; Faridi et al. 2021; Pal and Yasar 2020; Westerkamp et al. 2020	System framework and business model paper	
	Ready-to-wear clothing	Bullón Pérez et al. 2020		
	Silk production and marketing	Sharma and Kalra 2021		
	Fashion and textiles and other sectors	Mazumdar et al. 2021		
		Guo et al. 2020	Model formulation paper	
		da Cruz and Cruz 2020	Position paper	
Lifecycle analysis	Footprint and emission tracking	Zhang et al. 2020; Teh et al. 2020	System framework and business model paper	**(1) Information transparency** The emissions and footprint data are shared to all relevant parties in the blockchain ledger, including stakeholders, firms, brands, the government and customers. **(2) Instant traceability** Blockchain records and tracks the precise emissions and footprint created in the production stages of a product, rather than an estimated amount. **(3) Speed and performance** The traditional process to calculate trading credits and record emissions can be time consuming. With the blockchain ledger, the emission data in different production stages can be traced and recorded efficiently.
		Fu et al. 2018; Khaqqi et al. 2018		
	Emission trading	Lam and Lei 2019	Case study	

Circular economy	Fast fashion Fashion and textiles	Wang et al. 2020 Huynh 2021; Rusinek et al. 2018	System framework and business model paper	**(1) Information transparency** The information are shared to all relevant parties in the blockchain ledger, including stakeholders, firms, brands and customers. **(2) Instant traceability** Blockchain records the provenance of products, including, but not limited to, the origins, ownership and repair history. This facilitates the process of product maintenance and recycling. It also provides proof of authenticity and also acts as an anti-counterfeiting tool for resale. **(3) Speed and performance** The information can be shown in near real time.
Workers' wellbeing and social welfare	Social welfare	Choi and Luo 2019	Model formulation paper	**(1) Information transparency** The data are shared to all the relevant parties in the blockchain ledger.
	Workers' wellbeing in fashion supply chain	Terra 2021	System framework thesis	**(2) Instant traceability** Each survey response is auditable and can be traced back to prove the validity. **(3) Data security** Blockchain optimizes data confidentiality to protect the identity of workers. Also, the data record in the blockchain is immutable which means data cannot be altered or manipulated.

TABLE 10.3
List of Blockchain Cases in Practice and Benefits of Blockchain Technology

Aspects	Case no.	Led by	Source	Benefits
Proof of sustainable origin	1	Provenance, Martine Jarlgaard and others	Steiner 2015; Chitrakorn 2018; Beckwith 2018	**(1) Information transparency** The data are shared to all relevant parties in the blockchain ledger, including stakeholders, firms, brands, the government and customers to track the product origin. **(2) Instant traceability** Blockchain documents the provenance of products from the raw material to finished product in store. It enhances the supply chain visibility.
	2	IOTA, ALYX and others	Maguire 2019; IOTA Foundation 2019	
	3	Textile Genesis, Lenzing and others	BOF STUDIO 2021; Lenzing May 2019; Lenzing Sep 2019; Lenzing 2020	
	4	Textile Genesis, fashion for good, Kering and others	Fashion for good 2020; Carr 2020	
	5	Textile Genesis and U.S. Cotton Trust Protocol	Tenn 2021	
	6	Bext360, fashion for good and others	Fashion for good 2019	
	7	VeChain and Arket	Vechain 101 2020	
	8	VeChain, COS and DNV.GL	Vechain 101 2020; DNV n.d.	
	9	IBM and Burberry	Lee 2020	
Lifecycle analysis	10	IBM, Covalent and LinuxOne	Roshitsh 2020; IBM NewsRoom 2021	**(1) Information transparency** The emissions and footprint data are shared to all relevant parties in the blockchain ledger. **(2) Instant traceability** Blockchain records and tracks the precise emissions and footprint created in the production stages of a product, rather than an estimated amount.

Circular economy	11	Everledger, MCQ and Temera	Mercer 2021; Degenhardt 2021	**(1) Information transparency** The information are shared to all relevant parties in the blockchain ledger, including stakeholders, firms, brands and customers. **(2) Instant traceability** Blockchain records the provenance of products, including, but not limited to, the origins, ownership and repair history. This facilitates the process of product maintenance and recycling. It also provides proof of authenticity and acts as an anti-counterfeiting tool for resale. **(3) Speed and performance** The information can be shown in near-real time. **(4) Data security** The customers' identity is anonymous in order to protect their privacy.
	12	ConsenSys, LVMH, Microsoft and others	McDowell 2021b	
	13	Arianne, LUKSO and others	O'Connor 2019; McDowell 2019a; McDowell 2021b;	
	14	EON, H&M, PVH, Target and others	McDowell 2019b	
	15	Queen of Raw	McDowell 2021a	
Worker wellbeing and social welfare	16	ConsenSys, Levi Strauss & Co, Microsoft and others	McDowell 2019a; New America Foundation 2021	**(1) Information transparency** The survey chart is shared to all relevant parties in the blockchain ledger, including stakeholders, management teams and factory workers. **(2) Instant traceability** Each survey response is auditable and can be traced back to prove the validity. **(3) Speed and performance** It greatly reduces the data process time from three months to near-real time. **(4) Data security** The platform protected the identity of workers who filled in the wellbeing survey. Also, the data recorded in the blockchain are immutable, which means data cannot be altered or manipulated.

10.4 CHALLENGES

Although academia and industry affirm the feasibility and potential of blockchain for sustainable fashion, the technology itself still poses challenges for application, including: (1) high cost, (2) security, (3) scalability and performance, (4) reliability and (5) technology adoption. Table 10.4 shows the academic papers that describe the challenges or barriers to blockchain adoption in sustainable fashion.

10.4.1 Lack of Resources

Blockchain technology is still an advanced technology with a complex network which may involve a huge amount of investment in human effort, computer power, technical expertise and consultant employment for development and system maintenance (Agrawal et al. 2018; Caldarelli et al. 2021). It might be hard for small- and medium-sized enterprises (SMEs) to implement this technology due to affordability and skill constraints. In addition, suppliers in less developing countries might lack the resources and capability to adopt blockchain to monitor the production process of the product lifecycle and well-being of employees (Kshetri 2018; Boucher et al. 2017). It is hard to actualize the final goal of traceability without their participation.

10.4.2 Security Issue

A smart contract is a business logic written in the computer programme with the function of automatic contract execution. Hacking is commonly focused on attacking smart contracts. If a bug is spotted by a hacker, the blockchain platform itself is

TABLE 10.4
Academic Papers That Describe the Challenges of Blockchain Adoption in Sustainable Fashion

Content Overview	Reference	Paper Type
Barriers to blockchain adoption for sustainable fashion	Caldarelli et al. 2021	Qualitative exploratory study
Comparison of Gen X and Gen Y consumers' perceptions of blockchain ecolabels	Navas et al. 2021	Behaviour study
Examine opportunities, major trends and challenges in various industrial sectors, including the fashion sector	Dutta et al. 2020	Review paper
Examine opportunities and challenges to incorporate blockchain in the fashion industry, including the aspect of sustainability	Tripathi et al. 2021	
Review and classify the literature related to sustainable supply chain management, including the challenges faced in the fashion sector	Paliwal et al. 2020	
Examine opportunities, major trends and challenges in green supply chain, including fashion sector	Kouhizadeh et al. 2018	

prone to cyberattack, resulting in a huge financial loss. In 2021, more than $4 billion worth of cryptocurrency was stolen by hackers, and smart contracts dealing with financial assets or trading transactions were one of the key areas to be attacked. Yahaya et al. (2020) suggested Oyente Tool, a specialized analysis tool for identifying potential security bugs in a smart contract for the blockchain-based energy trading system. Therefore, it is recommended that the developer of the transaction platform, especially involving smart contract trading transactions (i.e., emissions trading), uses a testing tool to eliminate risks.

10.4.3 Scalability and Performance

More participants in the blockchain network and greater product complexities result in more blocks and a longer time for block-building, verification and confirmation. If the number of transactions in a certain period of time exceeds the network capability, a performance bottleneck and latency will result (Wang et al. 2019). In result, the data cannot be updated in the ledger in real time, and this adversely affects the efficiency of the supply chain. Weber et al. (2016) suggested that private blockchain might be suitable for enterprises requiring high efficiency. On the other hand, Agrawal et al. (2018) believe that only some key information on the product lifecycle should be selected to trace; otherwise, it will be inefficient and costly if the whole product lifecycle is traced to achieve full supply chain visibility.

10.4.4 Reliability

Although the digital information in a blockchain network is immutable and tamper-proof, the virtual information does not always accurately represent the actual scenario of material flow and the physical product (Kshetri 2018; Bischoff and Seuring 2021; Apte and Petrovsky 2016; Wang et al. 2019). Take the digital passport case for fashion circularity as an example; falsification of a material source or product might occur when the product owner registers the product. To prevent this scenario, the Arianne Protocol team hired professionals to verify the actual product to make sure the information provided is correct before it is listed on the Arianne blockchain platform. Blockchain cannot fully replace auditing because the transaction records should be verified to ensure they are accurate and aligned with the physical scenario (Apte and Petrovsky 2016).

10.4.5 Technology Adoption

Kshetri (2018) reported that the Everledger founder and CEO Leanne Kemp took one and a half years to bring all the parties of the supply chain together for a blockchain platform. It is difficult for fashion brands to ask their supply chain partners to participate in the network because the partners may not be willing to share information such as production capacity and suppliers with the general public, which could hinder their positions in the market. Brands might need to negotiate with suppliers to deal with the confidential information to balance the interest of suppliers and transparency. In terms of consumer adoption, consumers might not be very familiar with blockchain technology. Some may have just heard of the word "Bitcoin" but not know that blockchain can be applied in other areas, such as the blockchain ecolabels on fashion garments. Navas et al. (2021) highlighted that Generation Y will be the early adopters of blockchain technology, recommending fashion companies utilize advertising campaigns to educate and introduce this new technology to the general public.

10.5 SUGGESTIONS AND RECOMMENDATIONS

Blockchain is still an emerging technology for fashion application, especially in the area of sustainability. Challenges still exist at this stage; however, they could be overcome when the technology becomes mature in the near feature. The full potential of blockchain technology has yet to be discovered. SMEs that lack resources to build or manage their own blockchain system might consider subscribing to a third-party cloud-based blockchain platform. The service providers are responsible for the platform development and management, so the enterprises can focus more on their core business. Some blockchain service platforms (e.g., Hyperledger Cello) operate in a "plug-and-play" form with little programming effort (Hyperledger Foundation 2022). This user-friendly feature might be suitable for some SMEs to integrate blockchain into their business. Moreover, it is suggested to integrate blockchain ledger feature (e.g., provenance) as part of the existing transaction operations rather than build a completely new system for traceability. This might reduce the cost of development, shorten the learning curve for new technology, and improve operational efficiency. Furthermore, some fashion companies might have incorporated blockchain technology in their business operation, and it is recommended to further research on blockchain interoperability, which refers to cross-chain compatibility for the interaction of various siloed and fragmented blockchain networks (Jabbar et al. 2021). Cross-chain enables smooth information exchange and communication across network which has been found in finance and cryptocurrency sectors. For example, Skuchain, a blockchain platform for global trade, released the Digital Ledger Payment Commitment Corda Decentralized Application (DLPC CorDapp) to bridge the blockchain network of Corda and Hyperledger Fabric (Corda and Skuchain 2020). This enables transactions between the bank partners in the network. Fashion companies might take reference from other industries, leveraging the blockchain interoperability of various networks including sustainable origin, life cycle analysis, circularity and worker wellbeing, and finally create a complete picture of sustainable fashion.

ACKNOWLEDGEMENTS

The work described in this chapter was supported by the Innovation and Technology Commission of Hong Kong under grant ITP/028/21TP.

REFERENCES

Agrawal, Tarun Kumar, Vijay Kumar, Rudrajeet Pal, Lichuan Wang, and Yan Chen. 2021. "Blockchain-based framework for supply chain traceability: A case example of textile and clothing industry". *Computers & Industrial Engineering* 154(2021): 107130. https://doi.org/10.1016/j.cie.2021.107130.

Agrawal, Tarun Kumar, Ajay Sharma, and Vijay Kumar. 2018. "Blockchain-based secured traceability system for textile and clothing supply chain." In *Fashion in 21st Century China*, 197–208. Singapore: Springer. https://doi.org/10.1007/978-981-13-0080-6_15.

Andolfatto, David. 2018. "Blockchain: What it is, what it does, and why you probably don't need one." *Federal Reserve Bank of St. Louis Review* 100 (2): 87–95.

Apte, Shireesh, and Nikolai Petrovsky. 2016. "Will blockchain technology revolutionize excipient supply chain management?." *Journal of Excipients and Food Chemicals* 7 (3): 910.

Arthur, Rachel. 2017. "From farm to finished garment: Blockchain is aiding this fashion collection with transparency." *Forbes*, May 10. Accessed September 30, 2021. www.forbes.com/sites/rachelarthur/2017/05/10/garment-blockchain-fashion-transparency/?sh=526877c574f3.

Beckwith, Charles. 2018. "Op-ed: Blockchains could upend the fashion business." *The Business of Fashion*. Accessed September 30, 2021. www.businessoffashion.com/opinions/technology/op-ed-blockchains-could-upend-the-fashion-business/.

Bischoff, Oliver, and Stefan Seuring. 2021. "Opportunities and limitations of public blockchain-based supply chain traceability." *Modern Supply Chain Research and Applications* 3 (3): 226–243. https://doi.org/10.1108/MSCRA-07-2021-0014.

BOF STUDIO. 2021. "At Lenzing, innovating traceability technology." *Business of Fashion*, June. Accessed September 30, 2021. www.businessoffashion.com/articles/news-analysis/at-lenzing-innovating-traceability-technology/.

Boucher, P. Susana Nascimento, and Mihalis Kritikos. 2017. "How blockchain technology could change our lives, European parliamentary research service." Accessed September 30, 2021. www.europarl.europa.eu/RegData/etudes/IDAN/2017/581948/EPRS_IDA(2017)581948_EN.pdf.

Bullón Pérez, Juan José, Araceli Queiruga-Dios, Víctor Gayoso Martínez, and Ángel Martín del Rey. 2020. "Traceability of ready-to-wear clothing through blockchain technology." *Sustainability* 12 (18): 7491.

Caldarelli, Giulio, Alessandro Zardini, and Cecilia Rossignoli. 2021. "Blockchain adoption in the fashion sustainable supply chain: Pragmatically addressing barriers." *Journal of Organizational Change Management* 34 (2): 507–524. https://doi.org/10.1108/JOCM-09-2020-0299.

Carey, Lindsey, and Marie-Cécile Cervellon. 2014. "Ethical fashion dimensions: Pictorial and auditory depictions through three cultural perspectives." *Journal of Fashion Marketing and Management: An International Journal* 18 (4): 483–506. https://doi.org/10.1108/JFMM-11-2012-0067.

Carr, Amanda. 2020. "New partnerships for CanopyStyle close the traceability gap." Accessed September 30, 2021. https://canopyplanet.org/new-partnerships-for-canopystyle-close-the-traceability-gap/.

Chitrakorn, Kati. 2018. "Can transparency solve the consumer trust deficit?" *The Business of Fashion*, December 10. Accessed September 30, 2021. www.businessoffashion.com/articles/sustainability/consumers-are-distrusting-transparency-matters-in-fashion.

Choi, Tsan-Ming, and Suyuan Luo. 2019. "Data quality challenges for sustainable fashion supply chain operations in emerging markets: Roles of blockchain, government sponsors and environment taxes." *Transportation Research Part E: Logistics and Transportation Review* 131: 139–152. https://doi.org/10.1016/j.tre.2019.09.019.

Corda and Skuchain. 2020. "Trade finance, meet interoperability?" *Corda*, March 24. Accessed January 20, 2020. www.corda.net/blog/co-authored-by-skuchain/.

Covalent. n.d. "Blockchain-backed traceability." *Covalent*. Accessed September 30, 2021. https://covalentfashion.com/product-timeline/?year=19&month=09&day=07&hour=11&minute=38&second=22.

da Cruz, António Miguel Rosado, and Estrela Ferreira Cruz. 2020. "Blockchain-based traceability platforms as a tool for sustainability." *In ICEIS* (2): 330–337. http://dx.doi.org/10.5220/0009463803300337.

Degenhardt, Patrick. 2021. "How fashion brands are taking advantage of blockchain apparel." Accessed September 30, 2021. https://2021web.everledger.io/how-fashion-brands-are-taking-advantage-of-blockchain-apparel/.

De Vries, Alex. 2018. "Bitcoin's growing energy problem." *Joule* 2 (5): 801–805. https://doi.org/10.1016/j.joule.2018.04.016.

DNV. n.d. "My Story™: A blockchain-powered digital assurance solution." Accessed September 30, 2021. www.dnv.com/services/my-story-a-blockchain-powered-digital-assurance-solution-141277.

Dutta, Pankaj, Tsan-Ming Choi, Surabhi Somani, and Richa Butala. 2020. "Blockchain technology in supply chain operations: Applications, challenges and research opportunities." *Transportation Research Part E: Logistics and Transportation Review* 142: 102067. https://doi.org/10.1016/j.tre.2020.102067.

Elmessiry, Magdi, and Adel Elmessiry. 2018. "Blockchain framework for textile supply chain management." *Multi-Agent-Based Simulation* 22: 213–227. https://doi.org/10.1007/978-3-319-94478-4_15.

Faridi, Muhammad Shakeel, Saqib Ali, Guihua Duan, and Guojun Wang. 2021. "Blockchain and Iot based textile manufacturing traceability system in industry 4.0." *Cyberspace Safety and Security*: 331–344. https://doi.org/10.1007/978-3-030-68851-6_24.

Fashion for Good. 2019 "Tracing organic cotton from farm to consumer." Accessed September 30, 2021. https://fashionforgood.com/wp-content/uploads/2019/12/Fashion-for-Good-Organic-Cotton-Traceability-Pilot-Report.pdf.

Fashion for Good. 2020. "Tracing sustainable viscose industry-wide implementation." Accessed September 30, 2021. https://reports.fashionforgood.com/wp-content/uploads/2021/06/VISCOSE-TRACEABILITY-PILOT.pdf.

Fu, Bailu, Zhan Shu, and Xiaogang Liu. 2018. "Blockchain enhanced emission trading framework in fashion apparel manufacturing industry." *Sustainability* 10(4): 1105. https://doi.org/10.3390/su10041105.

Guo, Shu, Xuting Sun, and Hugo K.S. Lam. 2020. "Applications of blockchain technology in sustainable fashion supply chains: Operational transparency and environmental efforts." *IEEE Transactions on Engineering Management* 1–17. https://doi.org/10.1109/TEM.2020.3034216.

Henninger, Claudia. 2015. "Traceability the new eco-label in the slow-fashion industry? Consumer perceptions and micro-organisations responses." *Sustainability* 7 (5): 6011–6032. https://doi.org/10.3390/su7056011.

Henninger, Claudia E., Panayiota J. Alevizou, and Caroline J. Oates. 2016. "What is sustainable fashion?" *Journal of Fashion Marketing and Management: An International Journal* 20 (4): 400–416. https://doi.org/10.1108/JFMM-07-2015-0052.

Howson, Peter, and Alex de Vries. 2022. "Preying on the poor? Opportunities and challenges for tackling the social and environmental threats of cryptocurrencies for vulnerable and low-income communities." *Energy Research & Social Science* 84 (2022): 102394. https://doi.org/10.1016/j.erss.2021.102394.

Huynh, Phuc Hong. 2021. "Enabling circular business models in the fashion industry: The role of digital innovation". *International Journal of Productivity and Performance Management* (ahead-of-print). https://doi.org/10.1108/IJPPM-12-2020-0683.

Hyperledger Foundation. 2022. "Hyperledger cello." Accessed January 20, 2022. www.hyperledger.org/use/cello.

IBM Newsroom. 2021. "Covalent taps IBM blockchain to help track the carbon impact of its AirCarbon®-based fashion goods." *IBM*. Accessed September 30, 2021. https://newsroom.ibm.com/2021-01-13-Covalent-Taps-IBM-Blockchain-to-Help-Track-the-Carbon-Impact-of-its-AirCarbon-based-Fashion-Goods.

IOTA Foundation. 2019. "Simply track the provenance and authenticity of your shirt." *IOTA Foundation Blog*, May 15. https://blog.iota.org/simply-track-the-provenance-and-authenticity-of-your-shirt-6c4a09509d5a/.

Ivanov, Dmitry, Alexander Tsipoulanidis, and Jörn Schönberger. 2019. "Digital supply chain, smart operations and industry 4.0". In *Springer Texts in Business and Economics*, 481–526. Cham: Springer. https://doi.org/10.1007/978-3-319-94313-8_16.

Jabbar, Sohail, Huw Lloyd, Mohammad Hammoudeh, Bamidele Adebisi, and Umar Raza. 2021. "Blockchain-enabled supply chain: Analysis, challenges, and future directions." *Multimedia Systems* 27 (4): 787–806. https://doi.org/10.1007/s00530-020-00687-0.

Kassab, Mohamad. 2021. "Exploring non-functional requirements for blockchain-oriented systems." 2021 IEEE 29th International Requirements Engineering Conference Workshops (REW), 216–210. https://doi.org/10.1109/REW53955.2021.00040.

Kassab, Mohamad, Joanna Defranco, Tarek Malas, Phillip Laplante, Giuseppe Destefanis, and Valdemar Vicente Graciano Neto. 2019. "Exploring research in blockchain for healthcare and a roadmap for the future." *IEEE Transactions on Emerging Topics in Computing* 9 (4): 1835–1852. https://doi.org/10.1109/TETC.2019.2936881.

Khaqqi, Khamila Nurul, Janusz J. Sikorski, Kunn Hadinoto, and Markus Kraft. 2018. "Incorporating seller/buyer reputation-based system in blockchain-enabled emission trading application." *Applied Energy* 209: 8–19. https://doi.org/10.1016/j.apenergy.2017.10.070.

Kotilevets, I.D., I.A. Ivanova, I.O. Romanov, S.G. Magomedov, V.V. Nikonov, and S.A. Pavelev. 2018. "Implementation of directed acyclic graph in blockchain network to improve security and speed of transactions." *IFAC-Papers OnLine* 51(30): 693–696. https://doi.org/10.1016/j.ifacol.2018.11.213.

Kouhizadeh, Mahtab, and Joseph Sarkis. 2018. "Blockchain practices, potentials, and perspectives in greening supply chains". *Sustainability* 10 (10): 3652. https://doi.org/10.3390/su10103652.

Kshetri, Nir. 2018. "1 blockchain's roles in meeting key supply chain management objectives". *International Journal of Information Management* 39: 80–89. https://doi.org/10.1016/j.ijinfomgt.2017.12.005.

Kumar, Vijay, Tarun Kumar Agrawal, Lichuan Wang, and Yan Chen. 2017. "Contribution of traceability towards attaining sustainability in the textile sector." *Textiles and Clothing Sustainability* 3 (1): 1–10. https://doi.org/10.1186/s40689-017-0027-8.

Lam, Oi Wa Amy, and Zhibin Lei. 2019. "Textile and apparel supply chain with distributed ledger technology (DLT)." 2019 20th IEEE International Conference on Mobile Data Management (MDM), 447–451. https://doi.org/10.1109/MDM.2019.000-4.

Lee, Adriana,. 2020. "Burberry, IBM interns collab on product-tracing system?" *WWD*, October 7. Accessed September 30, 2021. https://wwd.com/business-news/technology/burberry-ibm-product-tracing-prototype-sustainability-1234628134/.

Lenzing. 2019. "Lenzing presented first blockchain pilot project at Hong Kong Fashion Summit." September. Accessed September 30, 2021. www.lenzing.com/newsroom/press-releases/press-release/lenzing-presented-first-blockchain-pilot-project-a.

Lenzing. 2019. "Lenzing traces its fibers with blockchain technology." May. Accessed September 30, 2021. www.lenzing.com/newsroom/press-releases/press-release/lenzing-traces-its-fibers-with-blockchain-technolo.

Lenzing. 2020. "New level of transparency in the textile industry: Lenzing introduces blockchain-enabled traceability platform." Accessed September 30, 2021. www.lenzing.com/newsroom/press-releases/press-release/new-level-of-transparency-in-the-textile-industry-lenzing-introduces-blockchain-enabled-traceability-platform.

Maguire, Lucy. 2019. "Matthew Williams is using blockchain to tell Alyx's story." *Vogue Business*. Accessed September 30, 2021. www.voguebusiness.com/technology/1017-alyx-9sm-blockchain-matthew-williams#:~:text=As%20Alyx%20presents%20its%20Spring,the%20business%20case%20for%20blockchain.&text=Key%20takeaways%3A,its%20items%20to%20blockchain%20ledgers.

Mazumdar, Somnath, Thomas Jensen, Raghava Rao Mukkamala, Robert J. Kauffman, and Jan Damsgaard. 2021. "Do blockchain and IoT architecture create informedness to support provenance tracking in the product lifecycle?" In *Proceedings of the 54th Hawaii International Conference on System Sciences (HICSS)*, 1497–1506.

McDowell, Maghan. 2019a. "6 ways blockchain is changing luxury." *Vogue Business*, May 14. Accessed September 30, 2021. www.voguebusiness.com/technology/6-ways-blockchain-changing-luxury.

McDowell, Maghan. 2019b. "H&M, Microsoft, PVH Corp collaborate in circular fashion initiative." *Vogue Business*, July 16. Accessed September 30, 2021. www.voguebusiness.com/technology/circular-id-eon-sustainability-blockchain.

McDowell, Maghan. 2021a. "Startup spotlight: Queen of raw applies new technology to old fabrics." *Vogue Business*, April 13. Accessed September 30, 2021. www.voguebusiness.com/technology/startup-spotlight-queen-of-raw-applies-new-technology-to-old-fabrics#:~:text=Technology-,Startup%20spotlight%3A%20Queen%20of%20Raw%20applies%20new%20technology%20to%20old,and%20sellers%20of%20unused%20fabric.

McDowell, Maghan. 2021b. "The blockchain playbook: From LVMH's Aura to Arianee." *Vogue Business*, April 26. Accessed September 30, 2021. www.voguebusiness.com/technology/the-blockchain-playbook-from-lvmhs-aura-to-arianee.

Mercer, Louise. 2021. "Logging the benefits of blockchain fashion sustainability and traceability." Accessed September 30, 2021. https://2021web.everledger.io/logging-the-benefits-of-blockchain-fashion-sustainability-and-traceability/.

Mukendi, Amira, Iain Davies, Sarah Glozer, and Pierre Mcdonagh. 2020. "Sustainable fashion: Current and future research directions." *European Journal of Marketing* 54 (11): 2873–2909. https://doi.org/10.1108/EJM-02-2019-0132.

Navas, Rebekkah, Hyo Jung (Julie) Chang, Samina Khan, and Jo Woon Chong. 2021. "Sustainability transparency and trustworthiness of traditional and blockchain ecolabels: A comparison of generations X and Y consumers." *Sustainability* 13 (15): 8469. https://doi.org/10.3390/su13158469.

New America Foundation. n.d. "Project history." *Survey Assure*. Accessed December 21, 2021. https://newamericafoundation.github.io/digi_survey_assure/index.html.

Niinimäki, Kirsi. 2018. *Sustainable Fashion in a Circular Economy*. Espoo: Aalto University.

O'Connor, Tamison. 2019. "How luxury fashion learned to love the blockchain." *The Business of Fashion*, April 20. Accessed September 30, 2021. www.businessoffashion.com/articles/technology/how-luxury-fashion-learned-to-love-the-blockchain.

Pal, Kamalendu, and Ansar-Ul-Haque Yasar. 2021. "Internet of things and blockchain technology in apparel manufacturing supply chain data management." *Procedia Computer Science* 170(2020): 450–457. https://doi.org/10.1016/j.procs.2020.03.088.

Papahristou, Evridiki, and Nikolaos Bilalis. 2016. "A new sustainable product development model in apparel based on 3D technologies for virtual proper fit". *ICT with*

Intelligent Applications, 85–95. ICT with Intelligent Applications. https://doi.org/10.1007/978-3-319-32098-4_8.

Papahristou, Evridiki, and Nikolaos Bilalis. 2017. "Should the fashion industry confront the sustainability challenge with 3D prototyping technology." *International Journal of Sustainable Engineering* 10 (4–5): 207–214. https://doi.org/10.1080/19397038.2017.1348563.

Paliwal, Vineet, Shalini Chandra, and Suneel Sharma. 2020. "Blockchain technology for sustainable supply chain management: A systematic literature review and a classification framework." *Sustainability* 12 (18): 7638.

Rauchs, Michel, Andrew Glidden, Brian Gordon, Gina C. Pieters, Martino Recanatini, François Rostand, Kathryn Vagneur, and Bryan Zheng Zhang. 2018. "Distributed ledger technology systems: A conceptual framework". Cambridge Centre for Alternative Finance, Cambridge. http://dx.doi.org/10.2139/ssrn.3230013.

Reike, Denise, Walter J.V. Vermeulen, and Sjors Witjes. 2018. "The circular economy: New or refurbished as CE 3.0?—exploring controversies in the conceptualization of the circular economy through a focus on history and resource value retention options." *Resources, Conservation and Recycling* 135: 246–264. https://doi.org/10.1016/j.resconrec.2017.08.027.

Roshitsh, Kaley. 2020. "Exclusive: Luxury brand covalent captures carbon in air." *WWD*. Accessed September 30, 2021. https://wwd.com/fashion-news/fashion-features/covalent-luxury-captures-carbon-aircarbon-material-1234599222/.

Rusinek, Melissa J., Hao Zhang, and Nicole Radziwill. 2018. "Blockchain for a traceable, circular textile supply chain: A requirements approach." *Software Quality Professional* 21 (1).

Schinckus, Christophe. 2020. "The good, the bad and the ugly: An overview of the sustainability of blockchain technology." *Energy Research & Social Science* 69: 101614. https://doi.org/10.1016/j.erss.2020.101614.

Sharma, Abhilash, and Mala Kalra. 2021. "A blockchain based approach for improving transparency and traceability in silk production and marketing." Journal of Physics: Conference Series. 1998 (2021) 012013, IOP Publishing, doi:10.1088/1742-6596/1998/1/012013.

So, Daniel. 2020. "Alexander McQueen launches MCQ, a blockchain-powered creative platform." *Highsnobiety*, August 25. Accessed September 30, 2021. www.highsnobiety.com/p/alexander-mcqueen-launches-mcq/.

Sori, Amir Abbaszadeh, Mehdi Golsorkhtabaramiri, and Ali Abbaszadeh Sori. 2021. "Green efficiency for quality models in the field of cryptocurrency; IOTA green efficiency." In IEEE Green Technologies Conference (GreenTech). https://doi.org/10.1109/GreenTech48523.2021.00101.

Steiner, Jutta. 2015. "Op-ed | blockchain an bring transparency to supply chains." *The Business of Fashion*, June 19. Accessed September 30, 2021. www.businessoffashion.com/opinions/news-analysis/op-ed-blockchain-can-bring-transparency-to-supply-chains.

Sutherland, Brandon R. 2019. "Blockchain's first consensus implementation Is unsustainable." *Joule* 3 (4): 917–919. https://doi.org/10.1016/j.joule.2019.04.001.

Teh, David, Tehmina Khan, Brian Corbitt, and Chin Eang Ong. 2020. "Sustainability strategy and blockchain-enabled life cycle assessment: A focus on materials industry." *Environment Systems and Decisions* 40 (4): 605–622. https://doi.org/10.1007/s10669-020-09761-4.

Tenn, Memphis. 2021. "U.S. cotton trust protocol and TextileGenesis™ announce collaboration." Accessed September 30, 2021. https://trustuscotton.org/us-cotton-trust-protocol-textilegenesis-announce-collaboration/.

Terra, Ana Claudia Wierman. 2021. "Blockchain as a tool to improving work condition in the fashion supply chain." Master's dissertation, COPPEAD Graduate School of Business, Universidade Federal do Rio de Janeiro.

Tripathi, Gautami, Vandana Tripathi Nautiyal, Mohd Abdul Ahad, and Noushaba Feroz. 2021. "Blockchain Technology and Fashion Industry-Opportunities and Challenges." In: Panda, S.K., Jena, A.K., Swain, S.K., Satapathy, S.C. (eds) Blockchain Technology: Applications and Challenges. Intelligent Systems Reference Library, vol 203. Springer, Cham. https://doi.org/10.1007/978-3-030-69395-4_12.

Vechain101. 2020. "VeChain's partnership with H&M expands to high-end brand COS." Accessed September 30, 2021. https://vechain101.com/vechains-and-cos/.

Wang, Bill, Wen Luo, Abraham Zhang, Zonggui Tian, and Z. Li. 2020. "Blockchain-enabled circular supply chain management: A system architecture for fast fashion". *Computers in Industry* 123: 103324. https://doi.org/10.1016/j.compind.2020.103324.

Wang, Yingli, Jeong Hugh Han, and Paul Beynon-Davies. 2019. "Understanding blockchain technology for future supply chains: A systematic literature review and research agenda". *Supply Chain Management: An International Journal* 24 (1): 62–84. http://dx.doi.org/10.1108/SCM-03-2018-0148.

Weber, Ingo, Xiwei Xu, Régis Riveret, Guido Governatori, Alexander Ponomarev, and Jan Mendling. 2016. "Untrusted business process monitoring and execution using blockchain". *Multi-Agent-Based Simulation* 22: 329–347. https://doi.org/10.1007/978-3-319-45348-4_19.

Westerkamp, Martin, Friedhelm Victor, and Axel Kupper. 2020. "Tracing manufacturing processes using blockchain-based token compositions." *Digital Communications and Networks* 6(2): 167–176. https://doi.org/10.1016/j.dcan.2019.01.007.

Yahaya, Adamu Sani, Nadeem Javaid, Fahad A. Alzahrani, Amjad Rehman, Ibrar Ullah, Affaf Shahid, and Muhammad Shafiq. 2020. "Blockchain based sustainable local energy trading considering home energy management and demurrage mechanism." *Sustainability* 12 (8): 3385. https://doi.org/10.3390/su12083385.

Zhang, Abraham, Ray Y. Zhong, Muhammad Farooque, Kai Kang, and V.G. Venkatesh. 2020. "Blockchain-based life cycle assessment: An implementation framework and system architecture." *Resources, Conservation and Recycling* 152: 104512. https://doi.org/10.1016/j.resconrec.2019.104512.

Index

Printed in the United States
by Baker & Taylor Publisher Services

Printed in the United States
by Baker & Taylor Publisher Services